F O O D

COMBINING

&

DIGESTION

OTHER BOOKS

by Steve Meyerowitz

SPROUT IT! ONE WEEK FROM SEED TO SALAD

RECIPES FROM THE SPROUTMAN

WHEATGRASS: NATURE'S FINEST MEDICINE

JUICE FASTING AND DETOXIFICATION

Published by
The Sprout House, Inc.
PO Box 1100
Great Barrington, MA 01230
413-528-5200

F O O D

COMBINING

&

DIGESTION

by Steve Meyerowitz

ILLUSTRATIONS BY

by Rick Meyerowitz

Table Of

CONTENTS

INTRODUCTION

The Art Of Food Combining   1

THE LAWS OF FOOD COMBINING

You Are What You Eat   3

THE FIRST LAW - QUANTITY   5

Keeping A Diary   6
Overconsumption   7

THE SECOND LAW - FREQUENCY

The Schedule   9
Regularity   10
Same Foods   11
Other Cultures   12

THE THIRD LAW · EATING CONSCIOUSLY 15

The Nose Knows   16
What A Sight!   18
Taste Tells   18
Real Hunger   18
Full Of It--Enough Is Enough   19
Chipmunk's Anonymous   20
Speeding Ticket   21
Hello Gracie   21
Time and Place   22
Silencio   23
Chew It. Don't Eschew It   24
Food As Energy   25

THE FOURTH LAW · SEQUENCE 27

The Order   28
The Mono Diet   29
Liquids   30
Pureés   31
Soups   32
Fruits   33
Vegetables   33
Beans and Grains   36
Nuts and Seeds   37
Dairy   40
Eggs   41
Meat, Fish, Poultry   42

THE FIFTH LAW · CHEMISTRY

Protein   43
Combining Proteins   44
What Are The Protein Foods?   45
Hydrochloric Acid   46
Acid And Protein   47

Starch   48
Cellulose   49
The Starch Foods   50
Starch and Protein   51
Acid And Starch   51
Sugar   52
Eating Sugar And Starch   52
Protein And Sugar   53
Fats   56
Oil And Protein Or Starch   58

THE FIVE FOOD GROUPS CHART

Proteins   54
Fats   54
Nuts   54
Sugars   54
Starches   55
Grains   55
Fruits   55
Sub-Acid   55
Melons   55

OTHER FACTORS INFLUENCING DIGESTION

Water With Meals   59
Protein, Oil, Hot & Cold   60
Other Causes of Gas   63
Common Gas Producing & Allergic Foods   65
The Iron Stomach   66
The Peanut Butter & Jelly Sandwich   67
The Theory Of Relativity   67
Nature's Food Combining   68

A PERFECT DAY

Pre-Breakfast   70
Breakfast   70
Lunch   71
Before Dinner Snack   71
Dinner   71
Just Desserts   72

A DREADFUL DAY

Rise and Shine   73
Snack Time   73
Lunchtime   74
Pre-Dinner Snack   74
Dinner   74
Late Evening Snack   75

WAYS TO STRENGTHEN DIGESTION

Exercise   77
Breathe   79
Massage   79
Herbs   79
Juices   80
Enzyme Supplementation   80
Colon Drink & Cleanser   84
Be Happy   84
How To Control Appetite   85
Factors that Weaken Digestion   85
Parties   86
Business Lunch   86
The Six O'Clock Fight   87

INTERVIEW WITH SWAMI SPROUTANANDA

About Diet   89

About Diet 89
Ice Cream and Nirvana 89
Happiness and Food 90
Choosing A Diet 91
Getting Adequate Nutrition 92
Vegetarianism 94
Overeating 95

BIBLIOGRAPHY 97

RESOURCES 99

INDEX 101

INTRODUCTION

True happiness is impossible without true health and true health is impossible without a rigid control of the palate.

Mahatma Gandhi

The Art Of Food Combining

Most of us think of food combining as a science governed by laws of chemistry. Protein digestion requires an acid medium, starch digestion requires alkaline enzymes, sugars like lactose require specific enzymes like lactase, and so forth. If digestion was simply a matter of chemical combinations, we could satisfy the unanswered questions with a combining chart. Indeed, there are people who walk around with pocket food combining charts that they whip out in advance of every gulp. Although we must respect the science behind food combining and abide by the chemistry involved, there are other, non-scientific, factors that play a more significant role in our effort to achieve optimum digestion.

When it comes to good digestion everything matters. Lifestyle, emotions, attitude, timing, habits and environment all contribute to the *"art"* of food combining. The whole scenario of our daily lives figures into the efficiency of our stomachs. Digestion is not just getting food in and out. It is the absorption and assimilation of nutrients for the construction and repair of cells and the nourishment of the "whole" body. In one respect, this books plays the very pedestrian role of simply teaching proper eating habits. This involves not only what, where, how, when and why we eat, but also our consciousness during the process of nourishing ourselves.

Food encompasses our whole world, everything we do, everywhere we go, no matter what country or culture. Our social interactions revolve around questions of: *Did you eat? What do you want to do...Eat? What do you want to eat? Where shall we eat? When do you wish to eat?* If we stop eating, the world becomes vastly different. Sadly, many people would experience how empty their lives are. Others may not know what to do with all the extra free time. Try fasting sometime and watch the world change.

We are endowed with miraculous equipment which can digest even the worst chemical combinations if we do not over-extend ourselves. We need only to learn our limits and become more conscious of what we eat while we are eating it. Unfortunately, we spend so much time chasing dollars that we forget our bodies and lose touch with what we put in them. Indigestion is simply a red flag. It tells us to slow down and watch what we put inside. We need to treat our stomachs like we treat a baby. Be sensitive. Savor every bite--even a drink of water! Don't worry about counting enzymes or vitamins. Our body chemistry is far too complex for us to regulate from the outside. But we can make a difference by developing good eating habits and avoiding bad ones. For starters, we need to eat less and our meals need to be less complex. Animals eat one food at a time. Let your goal be to simplify your meals. Eat consciously, judiciously and apply common sense. Most of all, enjoy your food and be happy, then you will digest every bite.

Steve Meyerowitz

THE LAWS OF FOOD COMBINING

The act of breaking down and digesting foods frees the forces inherent in them, forces related to the various complexes of organs. Material nutrients are thus transformed into forces that nourish the nervous system and the brain...Food is a support that can be made use of only to the degree that the individual spirit actively transforms it.

Rudolf Hauschka

You Are What You Eat

Laws? The human body does not operate like a political system. Yet, it is a system governed by physical laws. When these laws or limitations are exceeded, problems do occur. Keep in mind, the penalties are not always instantly evident. Our current political system has a budget problem. The federal budget has not been balanced for years. Yet, we still go on. The effects of this violation of the law will be felt later on. Our economy will inevitably suffer. Whether there is depression, inflation, higher taxes, or unemployment, someone will pay.

In the same way, you may violate the "laws" of your physical system and not "pay" until later with irritable bowel, hypoglycemia, parasites, candida, eczema or other such conditions as the price. After age 60 you might have to deal with diverticulitis, hardening of the arteries or kidney problems. Some of us pay right away in the form of stomach ache, headache, flatulence, indigestion or diarrhea. Many people walk around with problems for years and do not know it. Others sense something is brewing, but doctors are unable to find anything. The patient is uncomfortable and frustrated. Often, the answer can be found in what we eat and how we eat it. Most of our problems are related to food in some way. It is unavoidable. We are what we eat. The medical profession is now beginning to recognize it. The *National Cancer Institute* is suggesting we eat high fiber foods and increase our consumption of fresh fruits and vegetables. Even tobacco consumption, which has been a known cause of cancer for over 30 years, is for the first time being restricted in public places. Whether you are 50 years old with a newly discovered ulcer, or 5 years old and getting rashes after eating citrus fruit, there are certain laws we must live by if we desire efficient digestion and optimum health.

THE FIRST LAW

QUANTITY

In general, mankind since the improvement of cookery, eats twice as much as nature requires.

Benjamin Franklin

Everybody has their limit. Our stomachs are merely physical systems. Like different brands of machines, they vary in size, amount of digestive juices available, strength and content of those juices and ability to replenish themselves. If you want to wash your rug, you don't use your home washer and dryer. You bring it to the commercial laundry where the machines are bigger and the detergents are stronger. Even among the home models, there are some that are smaller, which do not wash as well and take a long time to dry. These are the limitations of the machines and we would do well to think of our digestive systems as also having limitations.

Americans, as a society, do not look into their stomachs. Advertising fools us into believing we can consume anything. We yearn for a "whopper" of a meal. The jingles sing about a double cheeseburger with pickles, onions, tomatoes, cabbage, fries, worcestershire sauce on a sesame seed bun. It is almost too much to say, much less eat! And don't forget the large soda, malted or thick shake. Our desires are way ahead of our stomachs. The old aphorism "the eyes are bigger than the stomach" is embarrassingly true and our main problem. We can go a long way toward better digestion--and better health--by remembering that we have stomachs down there which can only do so much. Let us not dissociate ourselves from the romance of eating and the physical task resulting from it.

We cause our indigestion. How would it be if your city had a big parade, one so large that it took a week just to clean up. The townspeople were distraught because it prevented business from operating normally in its aftermath and was beyond the city's capacity. In the same way, you might yell at your teen-ager because they made a mess in their room that was impossible to clean up without help. Or you might be spending more with your credit cards than you can afford to pay at the end of the month. Admittedly, we live in a deficit society. But you cannot blame society for your indigestion.

We bite off more than we can chew. If we want optimum digestion and superior health, we must practice self-control. Self-control begins in the mind. We need to make a conscious decision to monitor our food intake at every meal. In other words--discipline.

Keeping A Diary

Just as the decision to be a healthy individual requires a commitment to good health, the road to a stronger digestive system starts with ones thoughts about eating. Clarify your intentions. How do you do that? One way is to start a diary. Monitor yourself and your eating habits. Judge for yourself what you are doing that is bad for you. Write it down in a diary. Watch when you do it. Establish the patterns. You will be amazed at how overpowering the temptation of food is--so much so that you find yourself continually seduced into habits that are self-destructive. Again, it is discipline that is going to be your salvation. Scrutinize yourself with your diary. Reread it. Study it and then write down your intentions. They will be self-evident.

Perhaps you find yourself eating with a voracious appetite and with desperation in the way you consume your food. Then ask yourself, *why*? Were you deprived of food when you were a child? Is that over with now? Why do you still carry that habit? Perhaps you have a habit of finishing everything on your plate. You may have been taught as a child to finish your food. Maybe you were not allowed to go play if you didn't. But those once seemingly useful habits may backfire on you today because they simply do not work in your favor now.

The solution for you is to become conscious of what you are doing and to take smaller portions. Perhaps you did not feel adequately loved as a child and compensate for it in your eating habits. You shower yourself with a variety of tantalizing and sumptuous dishes which reward you with the pleasure that familial love never did. If these situations are no longer current in your life, you must learn to break these old habits. It is discipline that is going to redeem you. Scrutinize yourself using your diary. Study it and then write down your intentions. Even meditate on them or create daily affirmations to keep you in line with your purpose.

Overconsumption

Overconsumption is the number one cause of indigestion. America as a society epitomizes this. In Europe, Americans are immediately singled out as "fat," "dumpy," and "overweight." At home, antacids, laxatives, and headache remedies--all needed because of overeating--are the number one pharmaceutical sellers. Names like Alka-Seltzer, Rolaids, Tums are synonymous with America. We have become a Di-Gel society where the only thing actually being well digested is what plays on TV.

The diary method is great for working on discipline and overconsumption and for keeping your weight problem to yourself. But for those of you who are bold enough, you can go a step further and tell your friends. Friends can serve the same role as psychiatrists. Of course, you can work with professional psychiatrists and counselors but friends have the advantage of being with you when you eat. They see eating to which you are blind. Good friends, if asked the right way, will sincerely want to help and will not laugh at your request. (Of course, you must choose the right friend.) Your friend is your eating partner and counselor and can relate to you as a caring observer. You need their third eye to tell you what you do just as actors need a director.

You may find that you load up your mouth with far more than you can chew at a given time. Incomplete chewing results in premature swallowing which causes many problems. Perhaps you are the type who chews and talks at the same time. This lets air into your stomach and again more swallowing of incompletely chewed food. Or perhaps you are a speed eater who consumes whole soyburgers in a single bite. Maybe you guzzle down lots of water with every gulp. Whatever your particular habits are, let your friend become your mirror. True, you may jeopardize the friendship when they realize what horrendous habits you have! But if you pick good friends, you will not want to lose them and will work even harder to correct your problems. This method helps externalize your bad habits and eliminates patterns of avoidance and rationalization. You will learn a lot about yourself as well as your dinner manners. Re-learning our relationship to food--what we eat and how we eat it--has a ripple effect throughout our entire behavior and affects our whole lifestyle. Be prepared for positive changes.

THE SECOND LAW

FREQUENCY

Gluttony is the source of all our infirmities and the fountain of all our diseases. As a lamp is choked by a superabunance of oil, and a fire extinguished by excess of fuel, so is the natural health of the body destroyed by intemperate diet.

Nathanial J. Burton

The Schedule

The second law of optimum digestion concerns how you fit eating into your daily schedule. Some people never give this any conscious thought. They merely eat when near food. This can, of course, lead to some obvious problems. The workaholic, never slows down to eat a meal. He/she only grabs one when he can, "on the run," leading ultimately to digestive distress. He may not even notice being hungry because he is so busy pursuing his work. Food is not important to him or his health and thus he is eventually bound to suffer from his negligence. On the other hand, the mother who has to take care of her baby and constantly walks in or near the kitchen might be tempted to feed herself all the time.

One of the worst offenses is eating late at night. People who do not eat properly during the day end up eating late at night. This is the only time they slow down long enough to enjoy a meal! In the case of the young mother at home, it could be her response when she is unable to fall asleep or her fear of sleeping because she might not hear the cries of her baby.

Some people unintentionally follow an imposed schedule because of the logistics of their daily lives. Their mealtimes may conform to their own work hours or the work schedules of other family members. These are fortunate events because, regardless of their origins, the effect is to regulate the mealtimes.

Regularity

The ideal way to eat, however, is to be as regular as possible in both timing and variety of food. Regularity is usually a term reserved for the process of elimination, but regular "in" means regular "out." The planets, the oceans, nature--all operate on rhythms and so does the human body! The individual who wakes at seven O'clock each day, has breakfast at nine, lunch at one, and dinner at six-thirty, is going to have a better functioning digestive system than one who is eating times are unstructured. Although most of us do not recognize it, we have a very prominent internal clock which helps us make it through the day. Yes, you can reset and adjust the clock but it should not be done frequently. Like Pavlov's dogs, we salivate for food when dinner time comes around. It becomes a ritual. Six p.m. is when everybody collects to eat. Everything else stops. The English stop for tea time, the Latins for siesta. Few people choose to ignore mealtimes and that is for the best, because the effect of external schedules has an internal physiological benefit as well. Your juices start to flow at dinner time and you *become* hungry. No matter what criticism you may have regarding the tradition of family mealtimes, they still have the result of enhancing our physical ability to receive a meal.

Same Foods

Regular use of the same foods is very helpful. True, a lot has been said about the benefits of an allergy-free diet that rotates foods. But regular use of the same foods does not oppose that. It merely maintains that eating a diet structured around certain foods, without too many invasions by exotic or unusual foods, can have a regulating effect on your digestive system. Ideally, a diet is structured in concentric circles with a group of primary foods, in the center, eaten every week and a group of secondary or accessory foods taken along with it. Some secondary foods are butter, miso paste, garlic, salt, olive oil, onions, etc.; foods you include in your diet regularly but are only used in a condiment capacity. In a ring around these foods is another group that may be chosen once or twice per month but are otherwise not regular. Perhaps you include foods like mushrooms and avocado in your diet but only once or twice per month. Beyond this group lies the culinary twilight zone where unusual and exotic foods (unusual to you, anyway) are taken only on special occasions such as weddings, trips, dinner at other people's homes, restaurants, etc. For example, you may never buy artichoke hearts or brussel sprouts or kiwi fruit but enjoy these at restaurants and special occasions.

This is not to imply that foods far from the central focus of your diet cannot be taken, but more digestive energy will be needed to process that meal. Also, radical shifts in diet disrupt your whole system and may result in the inability to sleep, irritability, headaches, fatigue, gas, heartburn, even colds. On the other hand, being regular in your diet, both in terms of timing and types of foods, creates regularity in the intestines. Regularity in means regularity out.

Other Cultures

If you examine other cultures you may or may not find better eating habits, but one thing for sure, you will find different
mealtimes. The Indians (Far Eastern) eat their largest meal at
noon or 1 p.m. and eat only very lightly at 6 p.m. Latin Americans also hold their biggest meal during the midday. The
"Siesta" is a mid-afternoon break usually from 1-3 p.m. This is
not comparable to our lunch hour but more equivalent to our
"dinner." They take their time and eat a lot. Most stores and
offices close. Even Western scientists see the virtues of taking
the largest meal midday because the largest number of calories
are burned and less are stored as fat at that time. The British
are famous for their "tea time" although the tradition is not as
universal today. They schedule their mealtimes early and take
them quite seriously. Most restaurants close by 10 -11 p.m. The
French, on the other hand, dine very late and, like Americans,
hold their largest meal at night. In Paris, like New York, one
can get a square meal at any hour of the night.

The habits of Indians (India) and Latin Americans are very
much in sync with the rhythm of the planets. Both cultures
point to the synchronistic relationship of the human body and
the heavenly bodies. Just as the tides of the ocean come and go
on a daily schedule, the digestive fluids in our bodies also follow a natural rhythm. Their ebb and flow is influenced by the
cycles of the Sun and Moon just as the menstrual fluids in a
woman's body are tuned to the monthly moon cycle.

What are the consequences for someone who refuses to
establish their own eating schedule? Initially, there may be
headache, stomach ache, flatulence, distension or fatigue. Later
on, these problems may become chronic. You are failing your
digestive system by not establishing a dietary rhythm. As much
as possible, take your meals regularly.

Perhaps the biggest curse of an ill-scheduled eater is the loss of digestive strength. At any given time, the supply of digestive enzymes is limited. After a meal, this supply is exhausted. Sufficient time must be allocated for the generation of new juices as well as for rest. A meal schedule, structured at 5 and 6 hour intervals, 8 a.m., 1 p.m. and 7 p.m. for example, is ideal. Small snacks of fruits or liquids in between meals will not usually interfere. But heavy meals eaten closely together create an overloading of the system and a backfiring of normal digestion. Food that rests in the stomach and intestines too long or that travels through too slowly ferments and putrefies affecting overall health and well being. The result is a digestive tract that behaves like a traffic jam where all major arteries in and out are jammed up.

THE THIRD LAW

Eating

CONSCIOUSLY

Take care of your intake, physical and mental. Be careful what goes in. Every country has its immigration office...Your body is your country and there are many ports of entry. You should put immigration officers everywhere.

Swami Satchidananda

Eating consciously does not simply mean the eater should remain awake, even though many people do fall asleep while eating. It simply points to the bad habit we have as a society of being disengaged while we eat. To say we eat without thinking is an understatement. We focus on everything else but the food! We eat while we watch TV, read the newspaper, drive, talk, work, walk, plane, train, elevator and ride the subway. Although unappetizing to consider, there are those of us who even eat on their way to the toilet! In our third law for improving digestion, we strive to remain aware of what we eat and how we feel while we eat. Sounds simple, but it is the most violated of all the "laws" we will discuss. It is essentially the gastronomical equivalent of "being there."

There are several aspects to eating consciously. Perhaps most important of these is the "what, where and when." We must learn what the right foods to eat are and when and where it is best to eat them. Inconceivable as it may sound, we all find ourselves, at some time, eating things we do not like. Even though we may realize it, we do it anyway. What does this mean? Have we have lost our senses? Probably not, because we know enough to realize it is wrong. But we are "unconscious." In this regard, nature gives us sufficient tools to determine what foods are right to eat.

The Nose Knows

The first tool of conscious eating is the nose. Many things have been said about this fleshy promontory. It is probably the only sense whose external organ has, for centuries, been maligned, teased, snubbed, powdered, disfigured, refigured and generally rubbed the wrong way. To say the least, its importance to our health and well being has gone unrecquited. But whatever your feelings about its appearance, its function is no less brilliant than its sister senses. Food is to a nose as nectar is to a bee. Walk past a pizza parlor for verification. The nose leads us to the right foods, like the the Sirens sweet song lured their sailors.

Nevertheless, the art of smelling is shunned by society. Dare you even think of lowering your nose to the dinner plate and the aristocracy would sooner chop it off than withstand the affront of a proximal proboscis. On the other hand, if you could wear your nose on your cuff it would probably be acceptable to snuff your dish...discreetly. Simply speaking, the nose's primary mission is that of gastronomical reconnaissance. And why not? A little preliminary surveillance is a matter-of-course for other endeavors. Even boy scouts probe their paths prior to advancing. Apologies are extended to Emily Post for any affront to her "manners of the table." But noses are conveniently located adjacent to the mouth for good reason. When it comes to food, the nose knows.

Noses will tell us if a food is attractive, repulsive, fresh, spoiled, sweet, heavy or light. Restaurant eating in America does not allow us to use our sense of smell before we order. We are forced to order from a written description without the benefit of sight or smell. (Some restaurants do display desserts for visual evaluation.) At the grocers, however, we can see, smell and feel the prospective produce. But be careful the grocer does not catch you sniffing! If it is not right to eat, you'll know it. Even perfectly fine foods with healthy aromas may get a "no" if they are not right for the moment. The nose telegraphs the essence of the food to the nervous system which will either support the food's digestion or reject it.

**The nose properly proportioned on the body
in relation to its gastronomical importance**.

What A Sight!

Our second line of detection apparatus is sight. The sight of foods is sometimes the only available method of determination. Still, it provides us with much of the information we need to make an educated decision. The sight of a bowl of luscious deep blue concord grapes or a beautiful salad platter or an apple hanging from its stem is enough to whet your appetite. It is unlikely to fail you. Fresh fruits and vegetables are the easiest to discern on sight whereas prepared foods such as baked vegetable pies, casseroles, soups, quiches, sautes, etc. require smell or taste to help make the right selection.

Taste Tells

Where an aroma is not available and sight is not enough, the tongue is our final line of defense. Covered with hundreds of tastebuds, this field of tonsillar sensors reveals all. When in doubt, taste it. Taste offers the broadest range of input of any of the senses from salty to sour, hot to cold, sweet to bitter, mild to spicy, rich to light, etc. Its primary use is at home where, in your own kitchen, you can adjust the foods you prepare according to taste. It can make the difference between the sensational enjoyment of a dish to unspoken acquiescence (just another meal). In your home you can elect to turn down a dish if it does not taste right and create alternatives. When you are trying to achieve optimum digestion or when digestion is weak, eating at home is necessary to provide the greatest possible flexibility.

Real Hunger

Conscious eating also requires that we know when to eat and when to stop. Ideally, the most digestive energy is available when food is taken at a time of genuine hunger. But eating when truly hungry is the exception in our society--not the rule.

When a farmer has worked 6 hours out in the field, he returns to the table with an appetite. He is physically hungry. His body has burned lots of calories and after a brief (but important) rest, he is ready to "dig in" to a well deserved meal. You can be sure that meal will be well digested. His digestive system is primed to receive the meal and processes it. On the other hand, a person who is not really hungry and eats because the group is eating, brings little enthusiasm and little digestive strength to his/her meal. The meal is likely to cause some form of indigestion.

No one can teach you to know when you are hungry. Hunger is part of your instinctual mechanism. You either experience it or you do not. The problem in modern American society is that the average American never experiences true hunger. Food is always around. We readily succumb to its temptation and would probably not know hunger if we felt it! Pay attention to your sense of hunger. Abide by it and your ability to digest and assimilate food will increase significantly by this alone.

Full Of It--Enough Is Enough

Knowing when you are hungry is one thing; knowing when you are full is another. When it comes to our stomachs, we are like babies. Typically, we consume way over our capacity only to realize it later after the damage is done. Our minds seem to be on a 10 minute delay. Although our stomachs are full, we go on gobbling until later when we protest: *Oy, am I full!*

Holiday mealtimes are especially notorious for their participants' wobbly exits from the table. They stagger to the nearest sofa where they can rest and sometimes they fall asleep. One might think they just left the war zone, not the dinner table. Where was their awareness? The Romans ate with abandon and regularly regurgitated their excesses. But we are supposed to be civilized and beyond such gluttony.

Ideally, you should stop eating when your stomach is half full. This is a discipline which must be self-taught. You can have others help monitor you, but ultimately, it is an individual discipline. We can train ourselves to sense the level of fullness in our stomachs and you do not have to be "psychic" to do so. Our stomachs are only 24 inches below our heads; it is not so far to travel. Do not be discouraged if you fail to sense anything at first. You will improve with each meal. It is just a matter of consciousness.

Eating beyond the point of fullness is really overeating. The problem of overeating is finally coming into formal recognition and is being addressed by different groups who hold meetings and establish programs. The most famous of these is "Overeaters Anonymous." Here overeating is treated as an addiction just like any other form of compulsive behavior. People who make the effort to join such a group have developed a serious problem which is already at an extreme stage. The majority of us suffer from the same affliction but to a lesser degree so that the manifestations are more felt than seen.

Chipmunk's Anonymous

A major cause of overeating is the perverse habit of over filling the mouth. Many of us tend to use our mouths as a dumpster. We load it up to the maximum eating, chewing and frequently talking in the process. An overfull mouth forces quantities of food down the gullet before sufficiently chewed causing, at the very least, gas and distension. The emotional reflex causing this type of behavior is deprivation. We think we must take it all now for fear that the next opportunity may not come soon. You will recognize this type of gobbler because he resembles a chipmunk scurrying around with cheeks full seking to load in even more. Over filling the mouth results in incomplete mastication and quickly saps digestive strength. The solution: take in small amounts and do not add more until the first batch is fully chewed.

Speeding Ticket

Another part of this problem is eating too fast. We all have different rates of eating. Some can consume whole frankfurters in the blink of an eye. Others drag out a meal to the point where even their tea is sipped so exactingly it becomes a course in itself. Obviously, one must strike a balance, but the person with the better assimilation is the slow eater--by far. Part of eating consciously is pacing yourself. If you are guilty of gobbling, then monitor yourself. Chew more slowly. Make a deliberate attempt to stretch out your meal and avoid eating at times when you are in a rush. Food rushed down suffers from incomplete breakdown with results of flatulence, distension, poor absorption, vitamin deficiencies, irritable colon and nervous stomach.

Hello Gracie

One of the best ways to counter this, and other bad habits, is to say *grace*. Grace is an old world religious and family tradition wherein thanks is given prior to beginning a meal. This ritual has, for many, fallen victim to modern times, fast foods and changing values. You may feel it is not a part of your religion, but most religious traditions actually have some form of pre-meal ritual--meditation, prayer or thanksgiving. Even if you are an atheist, it is still a valuable practice to spend a few moments before the meal to quiet your mind and relax your body. Consider it a pre-meal orientation. *Grace* aids in the control of bad habits such as eating too fast, too much, overloading, incomplete mastication, etc. There does not have to be anything religious about it. It is just good practice. You may thank the earth for bringing forth her nourishing foods or thank the people who took part in its preparation. Or, you may thank the universe for allowing you, at this time and place, to nourish your body while others, less fortunate, cannot. If you are really an agnostic, just count to 25 and then start. The goal is to take a small amount of time to focus on the act of eating and relax the body. Take a deep breath...then start.

A SAMPLE GRACE

We are so very thankful for all this good food, for all this good food we have on our plates. We are so very thankful for the good company with whom we share this meal and for all the good friends and family we have in our lives. We are so very grateful for all that we have, for all the good things we have in our lives. May this food nourish and strengthen our bodies, purify our minds, add peace to our days and enliven our spirits.

Time and Place

Time and place are very important to eating, for eating is half ritual. It demands there be enough *time* for the "ceremony," else the after-effect is a sour stomach. So often, we eat while "on the go" or we "grab a bite." Sometimes you see people eating on the street or even in elevators and subways. And how many people do you know who are winners in the race for the one minute breakfast? If you extend this time-crunching mania to its extreme, you have a *meal in a tablet* or an *all-in-one* powdered drink. Such products exist already.

Ideally, eating is something you do sitting down. But this does not mean in a car or in front of the TV. To eat is to dine and therefore we have a "dining room" for its full enjoyment. What is your ideal eating environment? How does candlelight, plush carpeting, white dinner napkins, china and Mozart sound? Whether this is your style or not, it is perfect for digestion. Soft music goes great with a meal as does a peaceful ambiance. Your ideal eating environment, no matter what your style, should have a relaxed atmosphere. Even the family barbecue or picnic is great because the atmosphere is relaxed and peaceful, with plenty of fresh air (good for digestion) and good cheer. Take your stomach out to eat. Treat it like royalty and it will reward you with good feelings and bountiful health.

Silencio

Some religions practice silence with each meal. This relaxes the whole body and is ideal for focusing on the stomach and the act of eating. Silence affords the ultimate in food consciousness. There are no distractions, no chatter, no animated conversations. You are calm and your thoughts are free to settle on the food and the feelings that come up from it. Alas, most of us are not Yogis or Moslems and although silence may be helpful and therapeutic, especially for those with food additions, bad habits and digestive problems, it is not practical in our culture as a regular event.

Unpleasantries during mealtime can influence digestion.

The opposite of silence at the meal is, of course, boisterous conversation. Sure, everyone talks at their meals, but uncontrolled gabbing will turn any meal into a mess. Talking while you eat allows air to be swallowed with the food creating burping and/or hiccups. Loud, boisterous conversation at mealtimes is more widespread than anyone wants to admit. Many households use the mealtime to release their frustrations, anger, business and gossip. There can be arguments and even outright family fights. Some families turn the dinner table into a boxing ring. If this is the American way, then the religious approach is preferable, at least in terms of optimum digestion.

Chew It. Don't Eschew It

One of the benefits of a relaxed, consciously eaten meal is proper chewing. Chewing is the first stage of the digestive process for all solid foods. There are some rules of which you may have heard that suggest a particular number of chews such as 60 to 100 chews per mouthful. But, most people do not need to make a real count. The aim is simply to masticate the food into a bolus--a fluid mass--before swallowing. No solid pieces should be allowed to enter the stomach. Counting is good but a bit puerile for most adults. And if you lose your count, the proponents of this method make you start all over again. Use good sense. Chew your food into the tiniest pieces possible. This creates more surface area on which the stomach acids can act. Large, un-masticated pieces of food entering the stomach cause flatulence, distension and dyspepsia. It takes longer for the stomach to work on large pieces of food and uses up more enzymes. The simple, mechanical act of mastication can mean a world of difference to digestion and assimilation.

Food As Energy

The ultimate aim of good food awareness is to merge with your food. The goal of "grace" is to bring the food into harmony with your own vibration. Although the concept may sound very "Eastern," it is the basic intent behind saying grace. Food is a form of energy. We consume food to add to or maintain our own energy level. Everything, after all, can be defined as a form of energy vibrating at such high rates that we perceive it as solid. All foods, including raw and cooked vegetables, have a pulse which, beyond their vitamins, proteins and carbohydrates, influence us on a vibrational level.

This effect is not entirely unlike the excitement, calm or fervor we feel from different pieces of music. The music moves us, we say. Yet, we did not touch it or eat it. Sound does not have a physical form, yet we accept that it influences us. Its vibrational rates are slow and in a range that we can hear. But the vibrations of animate and inanimate objects are so high they are beyond the audible range. We can sense them, however, if we train ourselves to feel those frequencies.

You don't need to be a psychic or yogi to do this. Most of us sense these things and always have but are not conscious of it. The vibrations of different foods blend with our vibrations and influence us by charging, draining, balancing, calming or irritating us. Think of the last time you felt great or lousy after something you ate. You may feel lousy after eating a steak. But others might feel great. The effects are many and are apart from a food's nutritional content.

This is another reason why *grace* is so important. It allows you the opportunity to sense your response to the food before you bring it into your body. Your response will be either attraction, revulsion or neutral. In effect, you commune with your food. Whether it is McDonald's french fries or fresh concord grapes, you can harmonize with whatever is on your plate and bring it into your energy field before you touch it. If it is good for you, it will strengthen, relax or otherwise benefit you.

If you practice, you can learn to use your sixth sense to commune with your food, sense its vibrational effects and decide to accept that type of influence or not. You may decide to eat something else! If we feel scattered and the food we choose is centering, the food will make us feel more grounded. If we feel expansive and the food is energizing, we will feel more expansive. On the other hand, if we feel enervated and the food is sweet and light, we will feel more enervated. As an example, Macrobiotic concepts are structured around food as a form of energy. "Yin and Yang" refers to two types of fundamental energies which every food has to different degrees. Followers of Macrobiotic diets choose foods according to their known effects, of which all are based purely on vibrational levels. They have taken the trouble to chart all foods according to their vibrational effects or relative Ying and Yang-ness and have made this admittedly esoteric concept more available to the general public. It is just another way we can become more conscious of the food we eat.

THE FOURTH LAW

SEQUENCE

I saw few die of hunger. Of eating, a hundred thousand.

Benjamin Franklin

After deciding how much to eat and the right time to eat it and focusing in on some of your unconscious and troublesome habits, it is time to turn our attention to the hard reality of the dinner plate and the decisions about what to put on it. Our world of food is vast and modern lifestyles along with the development of the commercial food network has added new foods and new forms of old foods which, in their attempt to make eating more convenient, has frequently made digestion more difficult. We live in a "gourmet world" where food has become a product of fashion and a lifestyle that is more in tune with trends than nourishment. It is getting harder to find basic grains and beans in modern supermarkets and vegetables are becoming more imported, exotic, irradiated and even genetically altered. Decisions about what to put on our plates are getting harder and combining the right foods is less obvious with such a multiplicity of choices. But foods can still be ranked and categorized and the order in which you introduce different foods into your system during the course of a meal or series of meals, contributes enormously to successful digestion, assimilation and absorption.

There are different theories on how to organize a meal and the art of eating foods in proper sequence is, at best, controversial. But any effort to arrange your meal that contributes to the harmonious flow of the different foods through the digestive tract, is the right policy. A well ordered meal permits a smoother handling of your stomach's digestive chores with faster digestive times and better assimilation of nutrients. This means less gas, bloating, heartburn and other symptoms and a happier more energetic feeling after the meal is over.

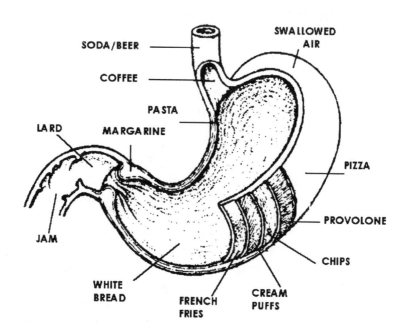

The Order

Our law of sequence rests on the theory that the easiest to digest foods should be taken in advance of the more complex foods. An apple, for example, should be allowed to go in and on its merry way without being obstructed by a veal parmesan. The aim is to keep a steady downward flow through the intestines avoiding any backlog or plugging up of the system. Trouble starts when foods from different groups, requiring their own individual chemical action and length of time, clash with foods from other groups requiring a different set of conditions.

You might call this the plumbing theory of digestion. Although our systems are not as simple as that, the laws of gravity and the mechanics of moving fluids and mass through pipes do apply. Foods can be categorized according to their different densities, water content and complexity in terms of fats, carbohydrates and proteins. The denser and more complex a food, the longer will be its journey through the processes of digestion, absorption and assimilation. Mixing foods of different categories and different densities complicates digestion and slows the whole system.

The Mono Diet

The ideal meal in terms of digestion is made up of one food. A breakfast of oranges is one from which you will never burp! Of course, too many oranges would violate the law of quantity, but generally speaking digestion will be quick, easy and efficient. Mixing oranges with grapefruits would also digest easily because they are from the same family (fruit) and same subdivision (citrus). Eating these foods is like eating a monodiet. The same thing can be said for apples and pears. These two foods can be considered as one so far as our stomachs are concerned because they are the same type. This is also true for peaches and plums, cantaloupe and honeydew, raisins and currents and all the berries.

Of course, all these foods belong to the larger family of fruits and generally any combination within the same family is considered good. But nature has its exceptions. Each food group differs in taste, texture, nutrition and thus the way it is treated by our digestive systems. Only the pairs we mentioned, those of the same subdivisions, can be considered excellent combinations. Salad greens also make excellent combinations. Boston lettuce, bibb lettuce, endive and romaine are all treated as one food as far as digestion goes. Even the addition of spinach, parsley or dill is excellent because they are in the same category--green leafy vegetables. Again, although all vegetables are good combinations, only those of the same family are considered an *excellent* combination. Some other examples of same families are a) beans and peas, b) walnuts and pecans, c) sunflower and pumpkin seeds, d) all bean sprouts, and e) all green sprouts.

Liquids

According to our theory, the easiest to digest foods would be liquids. Water would be our number one easiest to digest food and in turn anything with a high water content would be easy to digest. This means fruit and vegetable juices, light vegetable broths and many soups.

But there are exceptions. Milk, for example, is a liquid, high in water content, but not easy to digest. It is a concentrated liquid containing lots of fat, protein and carbohydrate. It takes more time to digest because it requires protein and fat digesting enzymes and tends to coat the stomach and neutralize the stomach acid. If you are going to drink milk, do it on an empty stomach, because milk will slow down the digestion of anything taken with it.

Another problem with milk, and certain other foods, is food allergies. If you have an allergy to milk, it will be hard to digest for that reason alone. The same would apply to citrus fruit if you have a sensitivity to citrus or to soy milk if you have a sensitivity to soy. Many people have trouble digesting milk because they lack the enzyme lactase which digests the lactose in milk. This enzyme becomes less available as we grow older and some races have an inherent lack of it. Soy milk is high in protein and fat and is difficult to digest simply because it is so complex.

Don't be fooled just because something is a liquid. Everything is relative. Yes, soy milk is easier to digest than soybeans and milk is more digestible than cheese. But if you have sensitivities to these foods, they will be difficult for you in any form. Learn your limits. Test yourself by having only small amounts of these foods if you suspect a problem. Try different brands, dilute them with water or avoid them completely if there is a problem.

Remember, the easiest to digest liquids are the juices--apple, carrot, spinach, celery, etc. They spend no appreciable time in the stomach and can be taken even by those with weak digestion. Juices go right into the intestines and 95 plus percent of their nutrients are able to be absorbed and assimilated by everyone. If you are not digesting your food, get your vitamins from fresh fruit and vegetable juices. Herbal teas are also nutritive and healing as are vegetable broths. Broths can be home made or purchased in powdered form and are very nourishing and easy to digest.

Pureés

Liquids are not always what we think they are. A blend of bananas and apple juice does not make banana-apple juice. It makes a puree of bananas. Just because solids are reduced into liquid form does not mean they disappear. Yes, you are eating solid food, only in tiny pieces! After all, where does the banana go? That illustrates the difference between a blending and juicing. A blender purees or liquefies a solid food. A juicer extracts the water content from a fruit or vegetable and separates it from the pulp. The higher the solid content of a liquid, the harder it is to digest.

One popular milk substitute is almond milk. Almonds are pureed with water or apple juice and then strained. If you do not strain the milk, you will be drinking solid almonds liquefied into small pieces. If you use half a cup of almonds to make the milk and you drink the whole thing (unstrained), then you are consuming half a cup of almonds. Some strainers pass more solids than others. If you have weak digestion, nut milk is an excellent choice, but make sure you strain out as much of the solids as possible. Also, be aware that some foods like cashews, bananas and papayas do not strain at all. That is to say they do not have solids which separate from their juice. If you blend cashews for cashew milk (delicious) and pour it through a strainer, the entire contents will pass through the strainer leaving nothing behind. Similarly, bananas and mangos run through a juicer leaving no pulp behind. You are always consuming solids when you juice or strain these foods. Please enjoy them. They are wonderful. But do not be fooled into thinking cashew milk is easier to digest than cashews and do not assume that there is such a thing as mango or papaya juice just because you see it in a store. Read the ingredients and you will see they are purees of the fruit mixed with juice from apple, grape or other fruits.

Soups

Soups are most difficult to categorize because there are so many varieties. There are vast differences, for example, between a miso broth and a hearty pea soup. Beans are hard to digest and a thick lentil soup is like eating a plate of beans. Just because water is added does not make a soup easy to digest. Each soup has to be examined individually. Some may have a milk base, chicken, beef or vegetable. They may include additives such as, MSG, thickeners or artificial flavors. Many have flour added, so if you are allergic to wheat, you would have difficulty digesting that soup. The easiest type of soup to digest is the light vegetable soup or broth. They are also easy to make. Just blend up your favorite vegetables such as asparagus, spinach or broccoli, add spices, water, and simmer. Soups, juices and water are best to have at the beginning of the meal.

Fruits

Next up in our sequence of easiest to digest foods is fruits. Fruits are the closest solid foods to water. In fact, their water content can be as high as 90 percent. This is apparent when eating or juicing a watermelon. After juicing a huge slice of watermelon, you are left with only a few tablespoons of pulp demonstrating that most of the fruit was water! On the other extreme, a banana yields no water and is all pulp. Bananas are starchy fruit and unusual as far as fruits go.

Fruits such as apples, pears, plums, peaches, melons, cherries, berries and citrus generally spend under an hour in your stomach. Since they are mostly water and soft fiber, no heavy protein or starch digesting enzymes or strong acids are required. Since this would be done mostly in the stomach, the fruits move through the stomach quickly. But there are exceptions. Bananas, avocados and coconuts are high in fat, protein and carbohydrate and take longer. Dried fruits like raisins, figs, dates, etc. are very concentrated with high fiber and high sugar content. They differ enormously from fresh fruit with less than 10 percent water content. They also spend more time in the stomach--generally 45 minutes to an hour and a half depending on quantity. The latter two types of fruits do not reflect the common attributes of fruits and actually belong lower down on our list of digestible foods. But in general, common fruits should take between 30 and 60 minutes to pass through your stomach.

Vegetables

Some vegetables digest in as little time as it takes for fruits. Tomatoes and cucumbers digest quickly and in fact have been categorized as fruits because they contain seeds although they are commonly categorized as vegetables. One definition of a fruit is a food that contains its own seed. Green leafy vegetables have almost as high a water content as fruits and overall a simple salad can take as little as one hour to hour and a half in the stomach. Of course, this does not take into consideration the dressing. The types of dressings and how much you use of them can complicate the digestibility of a salad. A pure olive oil and lemon dressing, for example, will extend the digestion time of a salad simply because the oil coats the leaves and is harder to digest. Tahini dressing or creamy (milk based) dressings take even longer. Of course, fancy salads like Caesar Salad or Waldorf Salad can have lots of other things in them such as ham, croutons, eggs and bacon bits making these salads more complex. A bulgur wheat salad includes grains. But speaking strictly in terms of vegetables, a common green salad with a simple oil and vinegar (oil and lemon) dressing should take between one hour to an hour and a half in the stomach.

Leafy vegetables are the fastest digesting group in the vegetable family. As the vegetables get starchier, the time required to digest them increases. Broccoli, brussel sprouts, summer squash, asparagus and cauliflower, for example, are so starchy they are usually eaten cooked. This is in contrast to the leafy salad green which would be unthinkable to cook. Steaming broccoli softens the fiber, breaks down the starch and adds water to the vegetable making it easy to digest. Of course the art of cooking requires that you know how much to cook to achieve an easily digestible food which keeps plenty of texture and taste remaining.

Some vegetables can be eaten raw or cooked. Cabbage is frequently served raw in salad and cole slaw and has a lot of fiber. But it has been known to cause gas and thus steamed cabbage, as served in stuffed cabbage or goulash dishes, is much easier to digest. The whole cabbage family should be treated this way including kale, swiss chard, bok choy and collard greens, among others. Carrots and beets are also prepared cooked or raw but are more fibrous than starchy. They are very high in water content and very juicy, but their fiber can keep you chewing for hours. Because of their fiber, these vegetables take longer than green leafy vegetables to digest.

Potatoes are the starchiest members of the vegetable family and take approximately 2 hours to digest when baked. They include sweet potatoes, yams, red potatoes, Idaho's, other tubers like rutabagas, and squashes like acorn and butternut. Potatoes are so starchy and concentrated that they should be cooked in order to digest them. However, your method of cooking is what matters and may complicate digestion. Steamed potatoes are easier to digest than baked potatoes because the food is softened by the water. Some people prefer baked potatoes because nutrients are leached into the water during the steaming process. Foods prepared by deep frying such as french fries are by far the most difficult to digest as well as the most unhealthy. Stir frying is better because the oil does not completely impregnate the food as in deep frying, but steaming or baking is the best. Frying adds oils which in itself would complicate digestion enough but deep frying oil is commonly so denatured that it is beyond the point of digestibility at all! Fried oils are often used all day or reused from the previous day. Once they are heated beyond their smoking point, they become harder and harder to digest. In a way, it is a kind of nuclear meltdown in that the molecular structure collapses and welds together in different shapes forming aldehydes and peroxides--toxins that our bodies cannot digest.

Some exotic members of the vegetable family are mushrooms, seaweed and sprouts. Mushrooms are high in protein and are considered the "meat" of the vegetable kingdom. One and a half (1½) hours is the average time for shitake, oyster and common edible mushrooms to digest. Seaweed is the equivalent of green leafy vegetables grown in the ocean. They are long and leafy and even though they appear black, when you hold them up to the light, their green color is revealed. They come in other colors, too. Dulse, for example, is a hearty red color. But they are all comparable to earth grown leafy vegetables in terms of digestion, taking only slightly longer to digest. Sprouts such as alfalfa, buckwheat, sunflower, clover, radish and other green leafy sprouts (non-bean sprouts) are considered in the same family as green leafy vegetables. But because they are so young and have such soft fiber, they are even easier to digest and more comparable to fruits than vegetables. Allow approximately one hour digestion time. Generally, nothing in the vegetable kingdom eaten in moderation takes longer than 2 plus hours to digest.

Beans and Grains

Beans and Grains are the next hardest family of foods to digest. Both are composed predominantly of starch but are considered respectable protein foods. They even have a fair amount of essential oils. As a group, beans and grains can take between 2½ to 3½ hours to digest depending on the quantity.

Glutenous grains such as wheat, rye, barley and oats cause some digestive troubles for those who are allergic to gluten. Gluten is the white sticky protein that is responsible for holding bread together. Actually, it is such a good glue that it is an ingredient in wallpaper paste and Plaster of Paris. Whole grains are slightly easier to digest than flour products partly because the gluten is not kneaded as it is in the breadmaking process. Non-glutenous grains such as corn, rice, millet, buckwheat (kasha), amaranth and quinoa are generally easier to digest than the glutenous ones. Of these, millet and buckwheat are the lightest and least concentrated grains and the quickest to digest. Corn causes some allergies and is fairly tough. Amaranth and quinoa are the highest in protein of all grains and fairly easy to digest.

Sprouted wheat, which is for all practical purposes the only sproutable grain, is easier to digest than regular wheat and has less gluten. Sprouting transforms enough starch in the grain to make it edible raw albeit only in small quantities. Soft wheat, commonly used in pastries, has less gluten and less protein and is easier to digest than common bread wheat. Sprouted soft wheat is pre-digested enough to eat as a raw snack although, again, in limited quantities.

On average, beans take longer to digest than grains. Generally, beans have about 10% more protein. Soybeans can have as much as 40% protein whereas amaranth, quinoa and wheat only get as high as 20%. This gives beans their reputation as respectable sources of protein but they are still mostly starch. Peas are the easiest beans (legumes) to digest. When picked fresh or sprouted, they can be eaten raw in limited quantities. Lentils, mung and adzuki can also be eaten raw after sprouting. But chick peas (garbanzos) and soybeans should not be eaten raw even after sprouting. Although sprouting reduces the cooking time and breaks down starch and protein, the process is not 100%. Although you may certainly enjoy some of them as raw sprouts, only small amounts are possible. More than that, most people would experience some digestive disturbance.

Tofu and tempeh are two relatively new foods to American cuisine. Tofu is made from curdled soy milk. It is a lighter product and very versatile. If you have trouble digesting soy, this may be the one soy food you can consume. However, it is still a difficult food if you are allergic to soy. Allow 2 hours for tofu. Tempeh is a bacteria cultured food. The bacteria slightly pre-digest the beans, but it is still a highly concentrated dish and takes 2½ to 3½ hours.

Beans are, of course, infamous for causing gas. This is due to the presence of trypsin inhibitors and complex sugars. Cooking, sprouting and rinsing the beans reduces these gas causing factors. (For more on flatulence and beans *see p. 61.*)

Nuts and Seeds

Nuts and seeds can take longer to digest than grains and beans depending on the quantity consumed. These foods are definitely more complex than the others. Some nuts are 30% oil, 60% protein and 10% starch. They have more protein and fat (the two most complex and difficult food factors to digest) than grains and beans. Just rub a few pecans or brazil nuts in your hand to feel their oil. It soon becomes obvious that these concentrated foods must be eaten with respect.

Their digestion would not be a problem, however, if it were not for commercial conveniences. Nuts come with a protective covering. Their hard shells are almost impossible to remove without special implements. Consider them big "beware" signposts. Our shelling machines remove the shells from almonds, brazil nuts, peanuts and sunflower seeds with impunity. The problem is that the convenience of having shelled nuts creates the impression that they are meant to be eaten by the handful. If it was not for automatic shelling, a handful of sunflower seeds might take 10 minutes to eat! Nuts were never meant to be eaten in any other way except one at a time. But that isn't even the worst of it.

Soon after the invention of nut shellers came the nut grinders, thrusting into our bellies a new food--nut butter. Sunflower butter, almond butter, sesame butter, tahini, cashew butter and of course, the infamous peanut butter. Nut butters combine thousands of nuts or seeds into one jar making a concentrate of all the oils, proteins and starches. You can consume hundreds of them with every spread of your knife. Just because they are not visible does not mean they are not there. Peanut butter, the chief offender, has the added complication of not being a true nut. Technically, it is a legume, a pea with a nutty, meaty taste. In fact, you can roast green peas and chick peas (both legumes) and consume them like peanuts. Peanuts are very high in both protein and oil. Just look at all the jars of vegetable oil on the supermarket shelves. Many of them are from peanuts. Avocados and coconuts, two fruits we discussed earlier, are similar in make up to nuts and seeds and when consumed in equal quantity are equally hard to digest.

The hardest nuts to digest are the ones with the highest oil content such as macadamia, pine, brazil, pecan and walnut. Macadamia nuts, which come from Hawaii and Australia, win the prize for the nut with the highest fat content.

This much oil brings forth another problem--rancidity. Once they are shelled, nuts loose their natural protection from the elements and deteriorate upon exposure to heat, light and air. Often, this is visible. Sunflower seeds turn from steel gray to brown or have yellow around the edges. The change in color is oxidation, the same process that causes a sliced apple to turn from white to brown. Other nuts develop a bitter aftertaste and still others have an unappetizing smell. Not only does it ruin the nuts, it is also harmful to our health. Rancid oils are purported to produce by-products that are carcinogenic.

Sprouting plays a very limited role with regard to nuts. Most nuts require germination in the shell and have long gestation periods. Sunflower seeds and peanuts are really the only practical sprouters. Sunflowers sprout in 2 to 3 days and have a definite improvement in taste and digestibility. Peanut sprouts take over a week and taste like fresh peas.

Treat nuts with respect. Too many will give you a bellyache. When possible, eat them in the shell. They are freshest that way and thick shells reduce the temptation of overconsumption. Eat nuts with salads and non-starchy vegetables. When taken in moderation, nuts combine well with most fruits. Nuts are frequently served with dried fruits which help break up their concentration. Do not eat them with grains or beans or other starches. When nuts are prominent in a meal, digestion can take as much as 2½ to 3 hours in the stomach and 4 hours more in the small intestine.

Please do not shy away from nuts and seeds. They are wonderful sources of protein, essential oils and minerals, but only if you can digest them! Nature has wrapped them in a hard package because they are so special. You don't need to eat many. Treat nuts with respect and they will nourish you well.

Dairy

Milk products are even more complex than nuts and seeds in that they contain lots of carbohydrate as well as fat and protein. As mentioned earlier, milk has the particular ability to neutralize stomach acids, slowing down the digestion of everything else you eat with it. When added to a meal, it lengthens the time the meal stays in the stomach.

Milk products include a wide range of different foods. First there is yoghurt--the light, low fat, fermented milk that has been partially pre-digested by the action of friendly bacteria and enzymes. If there is any milk product you can tolerate, it should be yoghurt. Other cultured milk products are also relatively easy to digest: buttermilk, kefir, sour cream, cottage cheese, farmer cheese and ricotta. They all have in common a culturing, aging or souring process which helps break down their protein and carbohydrate. Soft cheeses are generally easier to digest than hard cheeses because of their high moisture content. Hard cheeses like swiss, cheddar, provolone, parmesan, etc. are highly concentrated in their fat and protein and can take up to 4 hours in the stomach and another 4 hours in the small intestine being broken down.

Other types of milk products can make a difference. Raw milk is unpasteurized and maintains its natural enzymes to aid the digestive process. Although raw milk is easier to digest, it is also harder to obtain. Goat milk products are lower in protein and fat and are closer in nutritional content to human milk than cows milk. Some popular goat milk products include feta cheese, yoghurt, cottage cheese and even ice cream. Goat milk contains no casein a protein found in cow's milk and cheese. Goats are also healthier animals. They almost never get cancer while nearly 25% of all cows go to market with cancer. Cows are also subject to a wide range of anti-biotic, anti-depressant and anti-sexual drugs. This raises issues of animal cruelty and threatens the health of those who consume meat. Unfortunately, goats produce less milk, thus their milk products are more expensive. Buffalo milk and cheeses are even higher in fat and protein than cow's milk. They are uncommon and hard to get. But if you find some, you'll need a strong stomach to digest buffalo milk.

Buffalo milk is hard to get, even for a buffalo.

Eggs

Eggs and milk are usually considered in the same category because they are both products of animals, although eggs are not flesh foods themselves. Eggs are also very high in protein, fat and carbohydrates as are some cheeses, but the proportion of these nutrients in eggs are more suited to human nutrition.

Depending on how they are prepared, eggs can take 2½ to 3 hours in the stomach and 4 hours in the small intestine. Cooking methods such as soft boiling or poaching make eggs easier to digest than other methods.

Meat, Fish, Poultry

Fish is probably the easiest to digest of the three flesh foods because of its low fat content. Poultry is next and meat is hardest of all. Fish can pass through your stomach in 3 hours whereas meats can take over 4 hours in the stomach and 6 to 7 in the small intestine. These complex, high protein, high fat, high fiber foods take the longest of all foods to digest. The time can be further complicated by the addition of other types of proteins or fats such as cheeses. A cheeseburger, for example, with its layer of cheese and bun adds more protein from the cheese and starch from the bun to the already complicated process of digesting beef. The more you mix proteins, carbohydrates and fats, then the more complex the meal becomes. For best digestion of an already complex food, have flesh foods with vegetables alone and avoid starches and other types of protein. Multiple proteins increase the stress on the digestive system and can lead to enervation, fatigue, flatulence and putrefaction.

THE FIFTH LAW

CHEMISTRY

*As houses well stored with provisions are likely to be
full of mice, so the bodies of those who eat much are full
of diseases.*

Diogenes

Finally, we come to what everyone thinks is the most important element in a discussion of proper food-combining--the
chemistry. The irony here is that, for most of us, chemistry is at
best a difficult subject. The effort to juggle foods, or add pills
and enzymes to improve digestion, eludes most of us. The laws
of chemistry as they relate to digestion must be evaluated as
thoroughly as our other "laws." To make them our foremost
concern, however, is impractical and, in fact, futile. For most of
us, applying the laws of chemistry to our meals, is like opening
up a Pandora's Box--a maze through which we get increasingly
lost and bewildered. Those who take this route only, without
considering the other laws, often end up abandoning the
project and rebelling against the whole concept of food combining. Our aim here is to respect the chemical guidelines we
are about to discuss with equal reverence to that of the other
four laws we have already set forth.

Protein

Protein has a reputation for being the most important nutri-
ent there is. This is somewhat overdone. True, its role in building and repairing cells is indisputable. But its reputation is
derived mostly from the mistaken belief that it is hard to find

and necessary in large quantities. This is an issue separate from food combining which we cannot take adequate time to address here. Briefly, the problem involves a) the numbers which tell us we need a certain amount of protein at a certain body weight and age; b) the impression that protein is only available in a limited number of foods; c) that meals must provide "complete" proteins and d) that a minimum number of grams must be consumed daily.

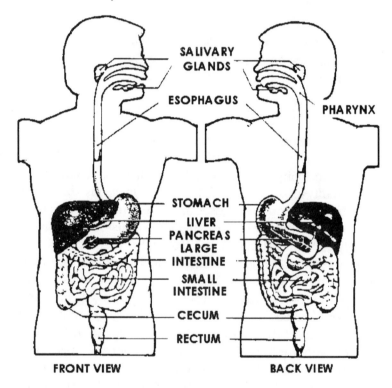

SALIVARY GLANDS
ESOPHAGUS
PHARYNX
STOMACH
LIVER
PANCREAS
LARGE INTESTINE
SMALL INTESTINE
CECUM
RECTUM
FRONT VIEW   BACK VIEW

Combining Proteins

This is not a food combining issue at all. It is a nutritional question and one that has been discussed for decades. It was initially brought into prominence by Frances Moore Lappe in her book *Diet for A Small Planet*. She expounded the theory that protein foods had to be combined daily in order to supply

the body with the right amount of the different protein components, namely amino acids. Not everyone agreed with her at
the time, and in the years since, the almost political controversy she initiated, ended with an event, rare in politics or lit-
erature: a retraction. In articles and books she has written
since, Lappe has aligned her thinking with the opposite camp--
those who believe that combining protein foods on a meal by
meal basis is simply an unnecessary effort.

What Are The Protein Foods?

Nuts, seeds, flesh foods (cow, chicken, fish), eggs and milk
are the main protein foods. In actuality, every food has protein.
It is required for growth. But the relative proportions of protein, starch and sugar determines whether or not it is a dominant element and thus considered a "protein" food. Some
foods, such as soybeans, stand out from other beans because
they have so much protein--approximately 35%. Most beans and
peas have between 20 and 25 percent protein which makes
them excellent sources of protein. But because they are also 60
to 65 percent starch, they can be treated as either starches or
proteins. Starches and proteins, as we will soon see, are a difficult food combination to digest, even when nature has created
the combination.

The highest protein foods are vegetarian. Spirulina and Chlorella, two ancient algae grown in high altitude lakes, are 65 to
70 percent protein. Nutritional yeast, a food manufactured by
friendly micro-organisms, is 50 percent protein as is Green
Magma, a powdered juice derived from young barley plants.
The above are all considered supplements and not primary
foods.

Other good protein foods are derived from grains and
beans. Tempeh and tofu are both made from soy and wheat
germ from wheat. Wheat has 15 to 20 percent protein at best,
but the germ holds most of that and when separated from the

grain yields more than 27% protein. Seaweed is also a surprise. Dulse has 25% in the dry form. Peanuts have about 26% protein (part nut, part bean), sunflower seeds 24%, almonds and pistachios 19%, pumpkin (and squash) seeds 29% and pignolia nuts 31%.

Hydrochloric Acid

All protein foods require an acid medium in the stomach in order to digest. Hydrochloric acid and the enzyme pepsin are the two primary digestive juices responsible for protein digestion. Protein takes a long time to digest. First there is mechanical breakdown by the teeth and the stomach muscles, then chemical breakdown in the stomach and intestines. The protein is actually dissolved by enzymes which break apart the molecular linkages yielding amino acids, the building blocks of protein. Many different enzymes are involved to work on the different sets of links. Finally, the amino acids are chemically small enough to pass through the walls of the small intestine and through capillaries into the bloodstream. The liver receives the amino acids through the portal vein and distributes them to tissues and cells. Your body synthesizes new proteins from the amino acid "pool" to build new cells, make muscles, repair tissue, form new enzymes, hormones and antibodies. In an emergency, the body can oxidize protein for energy but only when the supply of fats and carbohydrates are lacking.

True, if one amino acid is missing, your body cannot create a protein. This is the issue behind the whole controversy of protein complementation--the purported need to eat foods that contain enough "essential" amino acids to form protein. This is what Frances Moore Lappe first wrote about and what is meant by the "limiting" amino acid. Imagine making a barbecue for six friends with all the burgers (soyburgers), buns, ketchup, etc. in abundant supply and ready to go. Suddenly you realize you are out of charcoal and there is only enough to cook for two guests. The charcoal is the limiting factor. In the same way, your body will stop synthesizing protein when one amino acid runs short. That is why a diet balanced with different protein foods is more important than one that concentrates on a particular food however high in protein it may be. Spirulina is the highest of all protein foods, but it is not as good as getting a variety of proteins from different sources. The key issue here is that the amino acid "mix" need not come from one food or from one meal. The liver stores amino acids and distributes them to our system. As long as the general diet contains a mix of protein foods on a daily or weekly basis, the supply is not likely to fall short.

In terms of digestion, it is best to eat one protein food at a time whether complete or not. Mixing several different proteins in one meal is unnecessary and hard to handle. Too many protein foods cause an enervation of the system. You feel tired, even sleepy. Putrefaction can develop from undigested protein material and for the chronic abuser, this can mean the beginning of colon, liver, skin and other problems.

Acid And Protein

One might think that acid foods such as citrus fruits would contribute to the digestion of protein since proteins require an acid medium. However, to a certain extent, they actually limit the production of stomach acid by presenting your nervous system with mixed signals. When protein enters the mouth, the nerve endings on the tongue send signals to the stomach glands to secrete the appropriate juice. But the juices for digesting citrus are not the same as those for digesting meat. Eggs and orange juice are part of the all American breakfast, but they are not good stomach companions. There are many different kinds of acids and there is no evidence that fruit acids activate pepsin. In fact, they may destroy it.

Starch

Starches are the most common foods in our diet and although many starches might be considered "junk" food--potato chips, pretzels, crackers--they are more important than the name implies. Starch belongs to the larger family of carbohydrates, the most prevalent food in nature, which is simply a combination of carbon and water. All plants and grasses are carbohydrates. Also included in this family are sugar and cellulose. Starches are very complex with heavy chains of carbohydrate molecules requiring many steps before digestion. They are the best source of fuel we have for muscular activity.

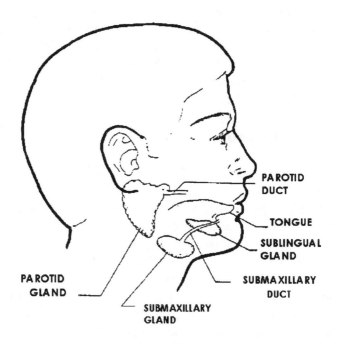

PAROTID DUCT

TONGUE

SUBLINGUAL GLAND

SUBMAXILLARY DUCT

PAROTID GLAND

SUBMAXILLARY GLAND

The Salivary Glands. Digestion starts in the mouth.

The first step in starch digestion begins in the mouth with the secretion of saliva. There are three salivary glands: the parotid above the jaw near the ears, the sublingual below the tongue and the submaxillary- below the jaw. Saliva contains the starch digesting enzyme amylase. But even if it contained nothing, it would still lubricate food during mastication. Without it, swallowing would be difficult. Saliva also makes the chemicals in food soluble to a point where the taste buds can detect the fundamental four flavors: bitter, sweet, salty and sour. Amylase contains the enzyme ptyalin, which begins dissolving the linkages in the carbohydrate chains into maltose and continues the process in the stomach. More amylase is secreted as necessary in the small intestine until the food is broken down into grain sugar--maltose. The by-products of maltose--glucose, fructose or galactose--are absorbed into the bloodstream. From there, glucose travels to the liver. The liver acts as a control center dispensing the fuel or storing it as necessary. Then, it is dispensed to the cells if the body needs energy. However, if there is no stress on the body, and you are a big starch eater, the liver will convert the glucose into glycogen for storage there or fat for storage in adipose tissue.

Cellulose

Cellulose is another form of carbohydrate, although an indigestible one. Ironic though it may seem, this indigestible food plays an important role in digestion. Cellulose, hemi-cellulose, lignin and pectins come from the fibrous portions of vegetables, the skins of fruits, the hulls of seeds and grains. Although they are broken by the muscular action of the stomach and softened by stomach acids, they are never dissolved. They are all indigestible in that we cannot break them down and absorb their sugar content. If we could, we would then be able to add a wide selection of plants and grasses to our menu similar to cows and horses. For us, the fiber keeps our starchy foods moving through the intestinal tract providing roughage and enabling the wavelike motion of the intestines known as peristalsis.

The Starch Foods

The common starch foods are grains and potatoes including:

Wheat	*Corn*
Rye	*Millet*
Oats	*Yams*
Barley	*Sweet Potatoes*
Kasha	*Squashes*

Also included are all foods derived from these staples such as:

Breads	*Pies*
Crackers	*Potato Chips*
Cookies	*Popcorn*
Cakes	*Corn Chips*
Cereals	*Baked Potato*
Pastas	*Mashed Potato*
Pretzels	

Squashes, such as acorn, butternut, buttercup, etc., are excellent starches, and vegetables such as cauliflower, broccoli and brussel sprouts are also starchy, although less so. Beans are all starchy even though they have respectable amounts of protein. They might include:

Lentil	*Kidney*
Navy	*Adzuki*
Black Bean	*Limas*
Garbanzo (Chick Pea)	*Green Peas*

Again, despite their high protein content, they are still 60 to 65 percent starch. As with wheat, foods derived from rice such as rice crackers, rice cereals, rice cookies, etc. are all starch foods as are those from millet, kasha (buckwheat), oats and barley. This, of course, also includes highly processed foods such as white bread and instant potatoes. Once the vitamins, fibers and enzymes have been removed, these foods are harder to digest since our bodies have to supply the missing nutrients in order to facilitate digestion. The only thing that is better about these foods is their shelf life.

Starch and Protein

Starch digestion begins in the mouth with the salivary enzyme ptyalin which envelops the food and continues the starch breakdown as it drops into the stomach. The more you chew these foods the better, since it increases the surface area for the ptyalin to act upon. Unfortunately, this enzymatic action stops if a protein food is eaten at the same time. Ptyalin and amylase are both alkaline enzymes and are neutralized by the hydrochloric acid in the stomach. If the starchy foods are eaten first and enough time is allowed for them to pass through the stomach, then protein foods can be eaten without interference. But if the combination of foods eaten are clearly starch and protein, then starch digestion will be slowed down until the protein food is digested. The starch then leaves the stomach in a semi-digested state and completes its digestion in the small intestine with additional secretions of amylase. Amylase is not secreted in the stomach.

Acid And Starch

For similar reasons, acid fruits taken with starch foods inhibit the digestion of the starch. Even one or two teaspoons of vinegar has enough acetic acid to suspend salivary digestion completely. This means tomatoes, berries, grapes, sour apples and citrus fruits will interfere with starch digestion. There is even some evidence that they not only inhibit ptyalin but are strong enough to destroy it. If you like orange juice with your toast in the morning, try having the O.J. ten minutes before and rinse your mouth before eating anything else.

Sugar

Sugar is the simplest type of carbohydrate. It is subdivided into three classifications--monosaccharides, disaccharides, and polysaccharides. Polysaccharides are the most complex form and the least sweet. Starches are considered polysaccharides. Monosaccharides, the simplest forms, cannot be broken down any further and are represented by glucose, fructose and galactose. Honey, milk and fruit contain monosaccharides. Glucose is the only kind of monosaccharide the body uses. Glucose is also called blood sugar or dextrose. In the liver, fructose (fruit sugar) and galactose (milk sugar) are converted to glucose for use by the body. Fructose is a main ingredient in table sugar and is sweeter than glucose.

Disaccharides contain two monosaccharides linked together. The most popular disaccharide is sucrose. Sucrose contains glucose and fructose and is the type of sugar in sugar cane, beets and maple syrup. This is the form of sugar commonly used in commercial foods from candies to soda pop. Too many sucrose foods drain your energy because they use up the B vitamins and minerals required to dissolve the sucrose link. Lactose, another disaccharide, is the hardest sugar of all to digest because it requires the enzyme lactase which, for most of us, is present only when we are children. Many people have allergies and indigestion when consuming milk products for this reason. Maltose is a disaccharide that is a by-product from the digestion of starch in grains and is now becoming popular in the form of barley malt and rice syrup.

Eating Sugar And Starch

Sugar, whether a monosaccharide or disaccharide, is one of the simplest foods to digest. Even water dissolves it. It requires only vitamins and enzymes and spends very little time in the stomach. Starch spends more time in the stomach, but the problem with combining the two foods begins in the mouth.

Starch depends on ptyalin in the saliva to start its digestion and sugar inhibits the secretion of ptyalin. The signals read by the nerve endings and taste buds in the tongue become confused in the presence of sugar. Thus even though you chew up and mechanically break down your bread or pasta, little or no chemical digestion takes place. The bread has to wait until it gets to the small intestine before amylase can complete the starch digestion process. This is a bad combination for the sugar food as well since its passage is slowed down by the presence of the starch food. The longer sugar stays in the stomach, the greater the chance fermentation will take place. Fermentation is the breakdown of sugar into alcohol and carbon dioxide. The carbon dioxide causes gas and distension and the alcohol robs the body of B-vitamins. Not only that but the sugar can no longer be utilized for its food value.

Protein And Sugar

Fermentation is also the problem with protein and sugar combinations. Again, the sugar foods, spend very little time in the stomach and when taken alone move along quite quickly to the small intestine. But protein foods can take hours. Rather than passing right through on its own time schedule, the fruit or sweet may remain for hours or more until the protein food digests. During this time they undergo fermentation in the warm, moist environment of the stomach. Ordinarily, the acids in the stomach would prevent fermentation but sugar has an inhibiting effect on the secretion of gastric juice. This is why eating sweets before a meal can spoil the appetite. However, if the meal is big enough so that a certain amount of gastric secretion takes place, the subsequent carbon dioxide released will drive stomach acid up the esophagus giving you that familiar warm feeling known as heartburn.

THE FIVE

PROTEINS

NUTS
MILK
PEANUTS
CHEESE

AVOCADOS
BEANS & PEAS
SOYBEANS
FLESH FOODS
OLIVES

FATS

Oils

OLIVE
SOY, CORN
AVOCADOS
SESAME
SUNFLOWER
ALMOND

All Nuts

PECANS
WALNUTS
MACADAMIA
PIGNOLIA
BRAZIL
ALMONDS

SATURATED FATS

COCONUTS
BUTTER
MARGARINE
FATTY MEATS

PALM KERN OIL
COCONUT OIL
CREAM

SUGARS

MILK
RICE SYRUP
MAPLE SYRUP
MOLASSES
SORGHUM

HONEY
BARLEY MALT
BROWN SUGAR
WHITE SUGAR
TURBINADO SUGAR

FOOD GROUPS

STARCHES

ALL GRAINS

SQUASHES
PEAS
ALL POTATOES
MOST BEANS

STARCHY VEGGIES

CAULIFLOWER
PEANUTS
BRUSSEL SPROUTS
CARROTS & BEETS

NON-STARCHY VEGETABLES

ALL LETTUCE	CABBAGE
SPROUTS (LEAFY)	CHINESE CABBAGE
SPINACH	CUCUMBER
SWISS CHARD	RHUBARB
SWEET PEPPERS	DANDELION
RADISH	GARLIC, ONION
SCALLIONS	BAMBOO SHOOTS
ENDIVE	CHIVES, LEEKS
CELERY	DILL, PARSLEY
CHICORY	ESCAROLE
WATERCRESS	KALE, COLLARDS

FRUITS

Acid Fruits	Sub-Acid	Sweet	Melons
ORANGES	APPLES	BANANAS	WATERMELON
GRAPEFRUITS	PEARS	DATES	HONEY DEW
PINEAPPLE	PEACHES	FIGS	CASSABA
TOMATO	APRICOTS	RAISINS	CANTALOUPE
LEMON	PLUMS	PRUNES	PAPAYA
LIME	MANGOS	PAPAYAS	CRENSHAW
SOUR APPLE	BERRIES	CURRENTS	
SOUR GRAPES	CHERRIES	DRIED PEARS	

Fats

Fats take the longest and are the most difficult of all nutrients to digest. However, you can help by selecting the finest oils and keeping your overall fat intake in balance within the rest of your diet. First let us define the terms. Fats are usually thought of as the white solid greasy part of animal foods and oils as the liquid squeezed from nuts, grains and seeds. The difference is that fats remain solid at room temperature and oils are liquid. Animal fats are mostly saturated while vegetable oils are unsaturated with a few exceptions. Both palm and coconut oils remain solid at room temperature. Saturation means that the molecules in the oil have been filled in with hydrogen atoms whereas in unsaturated fats some of the molecules are open. Liquid oils can be made solid through the process of hydrogenation which adds the hydrogen atoms artificially. The result is margarine, shortening and peanut butter that do not separate. Food manufacturers do this to increase shelf life, make the product creamier and resist rancidity. Unfortunately, hydrogenated oils are harder to digest. It takes more work to break down the bonds and helpful vitamins, minerals and essential fatty acids are destroyed in the hydrogenation process. Vegetables, grains, nuts and seeds have the highest proportion of unsaturated oils as do fish. Chicken and pork have a good amount of saturated fats. Beef has the most.

Do not avoid fats on the basis their difficulty to digest. Fats are very important nutrients. They provide more calories (heat energy) than any other food--9 calories per gram as compared with 4 calories from proteins and carbohydrates. They also help cushion the internal organs, line and protect the central nervous system, insulate us against heat loss, absorb and transport fat soluble vitamins and regulate fat metabolism. The most important members of the fat family are the polyunsaturates--linoleic, arachidonic and linolenic. They are known as essential fatty acids because they are not manufactured by the body and are necessary for normal cell growth and function. Sunflower, sesame, corn, soybean and safflower are excellent sources.

The fat content of your meal determines how long food will remain in your stomach. Fat digestion supersedes protein and carbohydrate digestion since enzymes must first work on the fat in order to separate the different nutrients. Normally, a meal with high fat content can stay in the stomach for 4 to 5 hours before it passes through to the duodenum. The duodenum is the first stage after the stomach and the passageway between the stomach and small intestine. The nerve endings here send signals to the gall bladder to send bile. Bile emulsifies the fat and starts the secretion of additional enzymes that further digest and separate it from the bile salts. The emulsified fats are now soluble enough to pass through the walls of the small intestine where they are carried by the portal circulation to the liver. The liver combines oils with protein forming lipo-proteins which it distributes to the cells and tissues.

In the stomach, fat digestion is assisted by acids, various enzymes, vitamins, minerals and phospholipids. Phospholipids, such as lecithin, are natural of many oils and foods. They help emulsify fats. Phospholipids are very valuable in the digestion of fats and are found in every cell especially the liver, brain and nervous system. The more unrefined your oil is, the better the chances it contains phospholipids and the valuable fat soluble vitamins like A and E. Soybeans are particularly rich sources of phospholipids. Unsaturated oils are easier to digest and absorb than saturated ones and thus it follows that vegetable and fish oils are easier to digest than animal fats. Avoid commercially processed oils as much as possible. They use dyes, caustic solvents, stabilizers, preservatives, heat and bleaches to extract and prepare them. In addition, the oil portion of a plant usually contains the largest percentage of oil soluble pesticide residues. Some say cottonseed oil even contains D.D.T. Avoid "cold processed" oils which can involve the use of chemical solvents even though there is no heat. Unrefined and cold pressed oils are the most nutritious and digestible. Remember that high heat makes even unsaturated oils less digestible and creates by-products which irritate the intestinal walls and interfere with good digestion.

Olive oil is the highest food in oleic acid with peanut and sesame oils next. Safflower oil is highest in linoleic acid with corn and soybean oils close behind. Other excellent sources of essential fatty acids, and relatively easy to digest, are almonds, pecans, sunflower seeds, olives, avocados, wheat germ, walnuts and pumpkin seeds. Olive oil and butter have the highest smoking points which means they are the safest for cooking. The smoking point is the point at which the oil breaks down. Butter does not create cholesterol as was once believed. Avocados, butter, soybean oil and the yolk of eggs are all excellent sources of phospholipids. All vegetable oils are not perfect. Palm kernel oil is a naturally hydrogenated vegetable oil used widely in confections because it creates a thick, creamy texture. Unfortunately, it is not easy to digest. Use of safflower oil is discouraged by practictioners of Ayurvedic medicine who claim it interrupts the assimilation of calcium and creates gallstones. Safflower is also a heavily sprayed commercial crop and, like cottonseed, is not regulated by the F.D.A. since the plant itself is not a food.

Oil And Protein Or Starch

Oil is similar to milk in that it slows the digestion of everything. It depresses appetite, inhibits stomach motility and delays gastric secretions. Creams, gravies, butters and oils taken in the same mouthful as nuts, cheeses, eggs or flesh foods create a multifaceted and extremely complicated meal. The same is true for fats and starch. The digestion of french fries is very slow and difficult because the potato is impregnated with oil. First the oil must be digested before starch digestion can begin. Lemon and green vegetables are about the only foods that are good combinations with oil. Green chlorophyll rich foods counteract the depressive effects of oils on the digestive process.

Other Factors Influencing

DIGESTION

Water With Meals

Although every issue in food combining raises questions, the issue of drinking water with meals stands out as one of the most controversial. It is really quite simple. The claim is that water dilutes stomach acids and enzymes thus prolonging the time in the stomach or flushing the food out of the stomach before it has been properly treated. Although this is accurate, it is not a simple black and white issue.

Water is an ingredient of many foods and for foods with a high water content, the addition of a reasonable amount of water is often harmless. If you have a glass of water after eating a watermelon, for example, it will make little difference. Having water with apples or peaches is not detrimental to their digestion since fruits spend little time in the stomach anyway and do not require hydrochloric acid. Hydrochloric acid assists only if there is lots of fiber that must be softened. Certain fruits like coconut and avocado are exceptions. Both contain low moisture, high fat, high protein and in the case of coconut, high fiber.

Salad greens, sprouts and green leafy vegetables of all kinds are all 80 to 90 percent water. Again, drinking water with these foods does not usually create a problem. High fiber vegetables, however, such as carrots, beets and cabbage, when taken raw, require a certain amount of stomach acid towards the end of the digestive cycle to soften the fiber. This fiber is not digested. Instead the cellulose acts primarily as roughage, although a certain amount of acid action can take place depending on the food. Even in those cases, the drinking of water only mildly effects digestion since it is not a digestible product. Starchier vegetables, such as broccoli, cauliflower and brussel sprouts, have increased their water content during the steaming (cooking) process. You can drink a fair amount of water with these foods as well, although less so than with leafy greens.

Dried fruits are concentrated high fiber, low moisture, high sugar foods. They do require some stomach acid to soften their fruit fiber. But water is also of value here. These fruits are so dry that the addition of water softens the fiber and acts as a lubricant and a solvent. Ideally, one would reconstitute dried fruits first before eating. Since this is not commonly practiced, the drinking of water helps reconstitute these fruits in the stomach.

Starchier vegetables such as baked potato also need lubrication which is usually done through the saliva or by mixture with other foods that have a higher water content. A salad for example, eaten with the potato helps keep the stomach lubricated. A meal of just potatoes and bread would be so dry that you would instinctively want to drink water. A judicious amount of water added here would function as a lubricant and help the stomach muscles churn and move the foods around. The key is allowing for normal motility of the stomach by either adding small amounts of water or mixing your drier, starchier foods with lubricating foods such as vegetables and salads.

Protein, Oil, Hot & Cold

When a generous amount of water is added to a protein meal, it dilutes the powerful stomach digestive juices: hydrochloric acid (HCL), rennin and pepsin. Water and oil rich foods such as cheese and avocados are also problematic, because they often contain protein and because oil and water do not mix. Drinking a bottle of soda pop with your burger either reduces the degree of digestion that takes place or prolongs the length of time in the stomach. Too much water taken on top of a full stomach can also act as a flush forcing some of the stomach contents to empty itself prematurely. Ice cold drinks chill the stomach and slow down digestion dropping the operating temperature of the stomach and halting stomach secretions. When this occurs, food stays in the stomach longer and extra energy is required to raise the operating temperature back to normal. The result is an enervation or (inner) exhaustion. Hot drinks also shock the system, although to a lesser extent. The closer the food or drink is to normal stomach temperatures, the less your digestive energy is drained.

Flatulence and Beans

Beans have acquired the stigma of flatulence foods. Gas is produced in beans because of the presence of 1) enzyme inhibitors and 2) bacteria and fermentation. Beans contain trypsin inhibitors that interfere with the action of trypsin. Trypsin is a digestive enzyme that is formed in the small intestine as a result of two other enzymes, enterokinase and trypsinogen, secreted by the pancreas. Trypsin breaks the peptide bonds in proteins particularly those associated with the amino acids arginine and lysine. Secondly, beans contain hard-to-break-down carbohydrates known as oligosaccharides. Oligosaccharides contain 2 to 8 sugars linked together. Raffinose is a trisaccharide containing 3 links of galactose, glucose and fructose. Stachyose is a tetrasaccharide (4 links) with an additional galactose. Carbohydrates with more than 8 links are known as polysaccharides. Their chemical bonds cannot be broken by enzymes in the stomach or small intestine. Instead, the job is accomplished by intestinal bacteria in the colon. Oligosaccharides are also found in cabbage family vegetables such as kale, collards, broccoli, brussel sprouts, cauliflower, turnips, rutabagas and kohlrabi to name a few, and in whole grains.

After they pass through the stomach and small intestine partially digested, anaerobic bacteria in the colon complete the digestion and produce by-products of carbon dioxide, hydrogen, nitrogen and methane. Foul odors are the result of relatively small amounts of hydrogen sulfide and ammonia among other noxious gases that result from the fermentation of incompletely digested food aging in the colon. Nitrogen and oxygen are, of course, the primary constituents of the air we breathe. They are present in flatus simply because some air is swallowed while eating and makes its way through the entire digestive tract before exiting. But swallowed air makes up only a small percentage of the flatulence problem. Intestinal gas is mostly hydrogen, carbon dioxide and methane.

Degassing the Beans

Rinsing the cooked beans and pouring out the cooking water helps eliminate the enzyme blockers since they are water soluble. The same goes for the oligosaccharides. Even frequent water changes is enough to gradually reduce these water soluble components. 1) Soak the beans for 3 hours. Discard the soak water. 2) Cook the beans in fresh water for 30 minutes. 3) Pour out the cook water, add fresh water and continue cooking. 4) Cook until soft. Pour out all or most of the cook water, add spices or other ingredients and serve.

Germinating the beans also converts and breaks down the oligosaccharides and trypsin inhibitors. The ideal approach would be to sprout the beans first and then proceed with steps 2 to 4 above. Sprouting also reduces the cooking time. Sprouting and rinsing together are the most effective way to maximize bean digestibility. Sprouting alone makes some small beans, such as lentils and mung, digestible raw depending on the number of days of growth and the amount consumed. The big beans always need additional cooking to make them fully digestible.

Fermenting foods into sauerkraut (cabbage), tempeh, tamari and miso (soybeans) creates cultures of active bacteria that also split the oligosaccharide bonds and convert trypsin inhibitors. One commercial product called Beano, contains the enzyme alpha-galactosidase which attacks the oligosaccharides. The enzyme is produced by the fungus aspergillus niger which is present in yeast, soy and other cultured and yoghurt-like foods.

Other Causes of Gas

Beans are not the only foods that cause gas. Milk and milk products, fiber foods and certain grains, are also guilty. In milk, sugar is again the culprit. The complex milk sugar lactose is too difficult for many people to digest because adults no longer produce "lactase," the enzyme that digests milk sugar. Undigested lactose often ends up in the colon fermenting and generating flatulence. Cow milk is really meant for cows, not humans. Even if you are young and loaded with lactase, cow milk contains protein and fat levels many times higher than human milk. Milk protein also requires the enzyme rennin, secreted by the stomach, to coagulate it. But rennin is absent from the gastric secretions of most adults. The manufacture of hard cheese and other solidified milk products intensifies the digestibility challenge. Goat milk and goat dairy products are much more digestible than cow products and closer to the protein and fat levels of human milk. If you must consume dairy products, emphasize yoghurt which is pre-digested by friendly bacteria. Because milk is often a "hidden" ingredient in other products such as desserts, cakes, puddings, cookies, etc., it could be the source of your digestive troubles without your awareness of it. Although, lactase supplement tablets are available, the ideal solution would be to leave the milk to cows and babies.

Certain grains such as wheat, oats, fiber and bran foods are also sometimes responsible for flatulence. Gluten, a difficult-to-digest protein in wheat, is frequently responsible for allergies in sensitive persons. The allergy is a response to undigested protein that acts as an irritant causing changes in the mucous membrane tract from the nose to the anus. The undigested gluten also ferments in the colon causing gas and distension. Gluten is actually two linked proteins glutenin and gliadin. It is so hard to break that it is used as an ingredient in glue and plaster of Paris. In bread, it becomes elastic and is the element that imparts body and strength to breads. Gluten is also a constituent of rye, oats and barley, although in smaller amounts. Fiber, which comes form the bran of these grains, is indigestible. It is actually the "shell" of the grain and passes through the digestive tract as bulk. Although it does not cause gas in itself, too much bulk causes a stress in the intestines that can aggravate the flatulence problem.

Insufficient hydrochloric acid (HCL) is another common cause of flatulence. Some individuals, especially the elderly, do not secrete enough stomach acid to digest their meals. Large pieces of undigested protein and fiber travel into the small intestine and colon where they begin to ferment. Individuals with low HCL production should eat smaller meals more often. Food combining laws should be strictly adhered to. Although HCL supplements help, increasing production naturally is the goal. Deep breathing exercises prior to the meal are helpful as are leg lifts and sit-ups which add tone to the intestinal area. Aerobic exercises that move the blood, the lymph and add oxygen and nutrients to the digestive organs are the ultimate solution. Fifteen minutes of aerobics, followed by a period of relaxation, deep breathing, and water in that order, is the best approach to increasing digestive strength. If HCL is the "fire in the furnace," then oxygen is necessary to fan the flames.

Supplemental Aids

The best short term remedy for flatulence is charcoal. Charcoal tablets are effective, efficient and non-toxic. They are nature's greatest absorbent of gas. A vegetable product, it is frequently made from coconut hulls. Tablets are superior to capsules because of their higher potency and greater purity. They are 100% compressed charcoal and eliminate the consumption of the gelatin capsule. There are few brands and they are sometimes hard-to-find. Requa brand is available in natural food stores. Dosage can be 2 to 6 tablets based on a 600 mg. dose. Start out with the small dosage until you have become familiar with its effects. Although charcoal is completely harmless and beneficial, it will 1) blacken the stool and 2) increase elimination. In addition to absorbing gases, it can also absorb nutrients. Use it on an empty stomach only once per day either in the morning or before bedtime. Avoid eating until one hour afterwards.

Drug store medications for flatulence often involve a chemical approach of reducing the size of the gas bubbles so they may pass through the digestive tract more efficiently. This, like many other modern medicines, addresses the symptom and not the origin of the problem. Short term use of the amino acid L-Histidine may be helpful in reducing the sulfur smell of odorous gas and stool. Use of chlorophyll supplementation with green foods such as blue green algae, spirulina, chlorella, alfalfa, wheatgrass and fresh greens, also reduces odors and lines the intestinal tract with this natural antiseptic and purifier. Garlic, either fresh if tolerable or in tablets, is another excellent antiseptic for the colon. A colon cleansing drink made of flax, psyllium or chia, taken daily, clears the colon of undigested debris and thus helps with colon management. *(For recipe see p. 84.)*

Common Gas Producing or Allergic Foods

Milk	Cabbage Family
Hard Cheese	Dried Fruit
Chocolate	Sweets
Wheat	Beans
Onions	

Avoidance of common allergic or gaseous foods is very important. It is unfortunate that beans are included on the list of gaseous foods. They are otherwise one of nature's most perfect foods and our best source of protein without excess fat. Do include them in your diet. Follow the recommended degassing method and enjoy them.

The colon is a two-way membrane which absorbs nutrients and oxygen but can also reabsorb poisons. You may have heard the expression: *the source of all disease begins in the colon.* Good colon health is fundamental to good health. Long term solutions to flatulence involve the abstinence from allergic foods, a close monitoring of the diet, exercises, supplementation and stress management. Since digestion involves so many essential organs and body processes, the true long-term solution is an overall health improvement and maintenance program.

The Iron Stomach

Our ability to handle difficult to digest foods varies from person to person and at different ages. Some people never have stomach problems. They may be active and in good shape, they may be unconcerned with their food or their bodies and they would probably never be reading this. Your friends may not worry about their stomachs but they may have back problems or skin problems or other concerns that you don't have to worry about. We are all endowed from birth with certain strengths and weaknesses. The fact that you are even interested in the subject of digestion signifies that this is an area you need to strengthen. The construction worker who gobbles down big whoppers and soda pop never gives his food a thought. In fact, he may think burping is healthy. But that is when he/she is young, strong and active.

But bad habits do catch up. Look at smoking. A smoker feels great after a cigarette. The smoking experience is pleasurable; it is only years later that smoking takes its toll on his body. It is now well known that smoking is the greatest single cause of lung cancer. However, we have not yet linked foods with specific health problems such as fried foods with colon cancer or a high sugar diet with cholesterol and heart disease. Nevertheless, it is a simple and unavoidable fact that wrong foods and chronic bad eating habits take a toll on our health, if not immediately, then years later. The friend with the iron stomach may get headaches or heart disease or high blood pressure or tumors or a stroke. Just because his food goes in and out does not mean it is being assimilated. If the mechanical process works, it just means he has good plumbing. But, the chemical breakdown of foods into nutrients and fuel, and their circulation to the cells and tissues, is the goal of good digestion. From this perspective, the person with the iron stomach has nothing about which to boast.

The Peanut Butter & Jelly Sandwich

If you were to follow the laws of chemistry assiduously, a peanut butter and jelly sandwich would be equivalent to a deadly weapon. Here is a food that breaks all the rules. Protein (peanut), starch (bread), sugar (jelly), fat (peanut) all mixed into one. In fact, to add to the mess, we are starting with an inherently complex food. The peanut, is high in starch, protein and fat which makes it difficult enough to digest on its own. Since this all American food was part of every child's upbringing, we might ask how did we ever survive!

The Theory Of Relativity

In the domain of food combining, most meals short of a mono-diet are problematic. The question is--problematic to what degree? Even the worst combinations can be properly digested if eaten in small enough amounts. If you have a bite of someone's peanut butter sandwich on an empty stomach, it is likely to have a negligible effect on your digestive system. If there is only a thin layer of peanut butter on your bread, then the bread becomes the dominant food. The stomach digests it first and works on the peanut butter later. This is analogous to a lamb in a bull pen. There is no question who is dominant. If there is a lot of peanut butter and a thin slice of bread, the stomach digests the peanut butter and the bread is finished after. Even if there is some fermentation, the amount of bread is relatively small and should not cause much discomfort. But if your sandwich has a lot of jam, a lot of peanut butter and a lot of bread and you eat a lot of it, then you have three bulls in the pen butting heads with a lot of rumbling and tumbling. Even if you escape the fight without too much discomfort, the meal is exhausting and everybody is somewhat bruised.

Nature's Food Combining

What about nature? Does not she create many foods, like peanuts which are combinations of protein and starch? If nature can do such a thing, why should we be so careful with our combinations? True, nature does create complex foods. But a peanut is recognized by your stomach as a single food. The stomach digests the starch first and then switches the enzyme secretions to digest the protein. This is a very different situation than having two distinctly different man-made foods which require different proportions of enzymes and timing. An example of a man-made food is a cookie. It is one thing to deal with natural grains, but when we grind grains, add sugar, baking soda, chocolate chips and yeast, we have a man-made food which is several steps removed from the original grain and infinitely more complex. From this, the stomach receives conflicting signals, the process becomes extended and ultimately digestion suffers.

Agreed. We all eat less than ideal food combinations, even if we know better. So what can we do about it? First of all, we can develop a little censor in us. The censor says: Buy peanut butter without sugar. Don't eat the fruit that is served with your salad. Leave over the baked potato that is on the plate with your steak. Don't buy cereal with added sugars. Push aside the orange juice that comes with your breakfast muffin. Choose the cookies that are made without nuts, etc., etc. In this way, you can control the degree of opposing foods. When you season your rice, use less butter or oil. If you want to melt cheese on your bread, use only a small slice of cheese. If you want raisins and nuts, mix two-thirds raisins and one-third nuts. Or, if you prefer, two-thirds nuts and one-third raisins. If you are at a smorgasbord and want to sample all the different foods, concentrate on the vegetables and starches, avoid the fruits and only dabble in the protein dishes. The toughest combinations are the ones with protein and starch, sugar and protein, fat and protein, and multiple types of proteins. All these require the production of lots of stomach acid and long term stomach transit times. Combinations involving sugars and starches are not as deleterious because their digestive effort is not nearly as extreme. Both foods depend relatively little on the stomach for their digestion. If you eat a bad combination with a fruit, the stress on the system is much less than if you were eating a steak.

A PERFECT DAY

This is the day you have been waiting for...a full day of perfect food combining. Let us start off with the easiest of all foods to digest---water. One can't get in too much trouble with water...unless it is polluted! Drinking water is the best way to start your day because it flushes the digestive system and gives you a clean start.

Pre-Breakfast

Now add liquid nourishment with a glass of fresh vegetable juice of carrots, beets, alfalfa sprouts, and apples. Oops! Did we mix a fruit with a vegetable? Yes! Because juices have no fiber content and spend little time in the stomach, the rules that govern the digestion of solid foods, frequently do not apply to the water extracts (juices) of those foods. Juices flow directly to the small intestine where they are easily absorbed. Allow at least 15 to 20 minutes for the juice to pass before eating breakfast.

Breakfast

This morning we will be serving brown rice cereal sweetened with rice malt. Rice malt is maltose, a disaccharide sugar derived from the digestion of rice. It is only mildly sweet. Certainly, if any sugar is going to be compatible with rice cereal, it is rice syrup. If we chew very well, we will ensalivate the rice thoroughly and minimize the inhibiting effect that malt has on the digestive enzyme ptyalin. You say you want to add raisins and bananas? True, they are delicious, but the more sweets you add, the more you complicate digestion. Okay, let's say you feel good and strong today, then just add a smattering of raisins and a few slices of banana.

Lunch

How about a sandwich? Choose the best bread you can find, preferably from stone ground organic grains with no added white flour or aluminum based baking powder. Now, add lots of sprouts or salad greens, some mustard, a slice of tomato, a pickle and a layer of mayonnaise with potato chips on the side. Uh oh...The food combining alarm started ringing off the hook! Hold the pickle, tomato and mayo. You cannot add an acid like pickle and tomato to a starch without interfering with starch digestion. Okay, we'll allow one slice of tomato since tomato is only mildly acidic...but perish the pickle! Pickles are cured in vinegar, a very strong acid even in small amounts. Mayonnaise is a protein that contains eggs plus saturated oil and vinegar. The addition of mayonnaise contributes several opposing combinations. If you do not want to give it up, smear it on thinly. Now for the potato chips. Isn't one starch at a meal enough? Truly, more than one starch is acceptable since starches are not that different or complex. But why must it be potato chips. Chips are deep fried in oil and are largely indigestible since the oil has permeated the potato thoroughly coating all the starch and prohibiting starch digestion for hours. Nix the potato chips, even if purchased from health food stores. Buy baked potato or corn chips or slice and bake your own. It's easy.

Before Dinner Snack

Keep it simple. An apple or a glass of herbal tea. Cinnamon tea is a wonderful digestive stimulant.

Dinner

Tonight's menu starts off with a glass of apple juice, a house salad and our main course: cubed tofu sauteed in olive oil, tamari, a touch of sesame oil, fresh grated ginger, garlic (or garlic sprouts) and sauteed snow pea pods (or Alaskan green pea sprouts), green peppers and mung bean sprouts. Alongside is a bowl of long grain brown rice. No dessert for now.

Magnifico. Here is a vegetable dish with one protein--tofu. True, oils are used. But olive oil is one of the most digestible and holds up well under cooking because of its high smoking point. The spot of dark sesame oil is added purely for flavor. Ginger stimulates digestion in addition to adding flavor. The vegetables all stand up well under light cooking adding texture and nutrition. The meal starts off with a juice cleansing the palate at the beginning of the meal when the most liquids should be consumed. Salad comes next in its proper position in the meal, ahead of the heavier foods. The dessert gets passed over.

Just Desserts

There is a time and place for everything. After the meal has had a chance to begin digesting, perhaps 1½ to 3 hours later, a modest dessert can be added. Before adding dessert, the dinner should have exited the stomach or nearly so. If you pick milk and cookies, you are mixing dairy, a starch and a sugar all in one. If you pick lemon-tofu pie, you have wheat in the crust (starch), tofu (protein), a sweetener (even if it is honey), and an acid fruit (lemon). Most desserts are complicated because they usually involve wheat, milk and a sweetener. The sweetener slows the digestion of the wheat and the milk slows the digestion of everything. If you just ate a date and had a cup of tea, it would be an excellent choice...at least in terms of food combining. If you choose to have a cookie or a piece of pie, have only small amounts. Take one or two cookies or have two bites of pie. You can handle difficult combinations as long as they are in manageable amounts. Remember, the bigger the meal, the less digestive power available for dessert.

A DREADFUL DAY

Rise and Shine

Start with two eggs, white toast, frosted flakes cold cereal, unfresh orange (the standard pasteurized kind) juice and hot coffee with cream and sugar. Eggs and toast are a protein and starch combination. The eggs take much longer to digest than the toast and so the toast will likely ferment. If you used jam, you would have further guaranteed the non-digestion of the bread by inhibiting its pre-digestion by ptyalin in the mouth. You will probably sip your orange juice throughout the first half of your meal. Thus, the acid from the juice will make it impossible for the bread to digest and scrambles the signals for stomach acid secretions necessary to digest the egg. The cereal is an extra unnecessary food and since the first one is not doing so well, it is not likely to fair much better. Coffee has four problems. 1) It is a liquid at the end of a protein meal thus diluting vital stomach acids. 2) It is an acid which confuses stomach acid production. 3) It contains milk which neutralizes stomach acids. 4) It is a hot stimulant and thus promotes a partial evacuation of the stomach contents before digestion is completed.

Snack Time

It is only 11 a.m. but the snack cart at the office is coming around with its sticky buns, chocolate filled croissants and sugar shellacked pastries and coffee. If you survived breakfast at all, you will need some of the sugar in this meal for energy and the coffee to keep you alert after the enervating morning meal.

Lunchtime

Today's special is a bacon, lettuce and tomato sandwich on white bread with melted American cheese and a cup of coffee. The lettuce is the best part of this meal. It combines properly with all the other elements except the coffee. The problem is, the rest of the items do not combine well with each other. Bacon is a protein and a fat and as such conflicts with the white bread (starch). The American cheese is another protein and another fat which further insures the indigestion of the bread and slows down the digestion of all other foods because it is a milk product as well as a fat. The tomato is relatively innocent in this crowd except that its mild acid would interfere with starch digestion in the mouth. Since the bread won't survive the trip anyway, it is immaterial. Each of these foods also carries a bag of inherent problems. The American Cheese is not really cheese at all but cheese "food" or cheese "spread" made from miscellaneous cheese parts which have been reprocessed. The white bread lacks any vitamins or minerals normal to bread which could aid digestion. Even the white iceberg lettuce, the most nutritious element of this meal, is the least nutritious of the lettuce family and is probably laden with pesticide residues. Of course, let us not forget the coffee and its usual side effects.

Pre-Dinner Snack

The snack cart returns with more goodies such as salted nuts and crackers plus red dyed pistachio nuts. Protein (nuts) and starch (crackers) make very poor digestive partners.

Dinner

At this point, not only is there little room left for a full dinner, but your supply of digestive fluids is more than likely exhausted. Nevertheless, dinner will not be pre-empted. First up: an alcoholic beverage such as beer or wine. Next an appetizer of deep fried mozzarella sticks, then house salad, buttered garlic bread and a main course of fettucini alfredo. Of course there is dessert, too...pecan pie, but this time no coffee, just a double espresso.

Alcohol robs the body of the B-vitamins necessary for digestion and stresses the liver which is needed during digestion. Deep fried *anything* is unhealthy, but in this case it is mozzarella--a hard cheese, fat/protein/dairy product--one of the hardest of all foods to digest. The fettucini alfredo is pasta (starch) with a cream sauce usually made from a soft ricotta cheese. Soft cheeses are lower in protein than hard cheeses but are still high in fat. This dish is very rich and loaded with opposing food combinations. Both the pasta and the cream sauce are equally dominant on the plate and the portions are typically large. The digestion will be slow and enervating. This is the kind of meal you will need to rest from afterwards but don't even think of lying down less you fall asleep. In this regard, the espresso is probably more a necessity than a choice. Pecan pie delivers yet another complicated and antagonistic combination--protein and starch in the form of pecans and wheat. To add further injury, pecans are extremely high in oil and the pie is high in sugar. The sugar ferments in the stomach. The pecans prevent the digestion of the crust and the oil from the nuts retards the whole process. This is a meal meant for heartburn!

Late Evening Snack

Time to go to bed, but first let us reward our stomachs for a long day's work. How about chocolate chip cookies, milk, candy bars or a soda? Since food does not digest in a supine position, the likelihood is that the snack will lay in the stomach all night. This gets the next day off to a bad start because the system is still clogged from the previous day. The best solution for ending this vicious cycle is a fast. Or, you could consider changing your dietary habits!

WAYS TO STRENGTHEN

DIGESTION

The ultimate challenge is to stop eating when you are half-full. It may sound crazy, but with this simple act alone, all your digestive troubles could be solved. -S.M.

So far, we have discussed methods of improving digestive efficiency. We have examined the types of foods we eat, watched our chemical combinations, the quantity and frequency of our meals and the time, place and other circumstances surrounding our meals. Now, we shall focus in on what non-food measures we can take to physically strengthen our digestive systems.

Exercise

Just as water puts out fire, air increases it. The power beyond our digestive systems is "fire." If this concept seems a little foreign to you, it is understandable. But that is the way it is referred to in Eastern medicine and philosophy. The stomach is the furnace of our bodies. Our food is the wood; digestion is combustion and the energy released keeps us healthy and alive. Anything you can do to increase your "firepower" will aid your digestion.

The specific exercises that increase oxygen to the lungs and the entire body are known as aerobic. Aerobic exercises include energetic walking, running, swimming, trampolining, jumping, skipping, hopping, dancing and cycling. Sports that include these activities such as football or baseball are also aerobic. Non-aerobic exercises, such as weight lifting or yoga, work on muscles and organs but do not leave you huffing and puffing. Golf is another example of a non-aerobic sport. Aerobics make a difference in your digestion because they increase your body's ability to take in oxygen and maintain an overall high oxygen level throughout your cells. All other things being equal, this means that calories are burned better, tissue cleansing is at an active rate and enzymes are in normal production. Most experts agree, if you exercise one half hour per day every day, you should see a measurable improvement in your overall health and thus your digestive power. On the other hand, if you only exercise one half hour per week, there is not likely to be any significant benefit.

Believe it or not, one of the best exercises for digestion is the trampoline. Unlike jogging which pounds away at the skeletal structure, trampolining is less "jogging" to the system. When trampolining, all the digestive organs get massaged. The intestines, liver, pancreas, stomach, etc. gently rise and fall without the same shock effect that takes place when your feet hit the pavement. If you think about it, most of the time our organs and glands remain in the same position opposing gravity when we stand and lay down. Trampolining enables them to change from their usual positions, free up tensions and release fluids. Hopping on a pogo-stick has the same benefit although it is less efficient. Swimming is also wonderful. The body constantly changes attitude and pitch and the organs gently flow into new positions. But if you cannot establish a regular exercise program, just walk. Walking after meals helps keep the blood circulating and the organs active. Sitting tends to cramp the organs and lying down after a meal tends to turn off or reduce the pace of digestion.

Breathe

If you cannot establish a daily exercise program, take a few minutes before each meal to breathe. Deep breathing brings fresh oxygen to the cells, awakens the nervous system and sends blood to the brain, lungs and stomach. Any deep breathing is fine as long as you do it for at least 5 minutes and then relax before you begin eating. Yoga practice includes many breathing exercises known as pranayama and they are all excellent. Yoga by the way, although it is not aerobic, includes many exercises which cause expansion and contraction of the stomach, liver and intestine areas. It can be extremely beneficial in stimulating and cleansing those organs. Yoga is known not as exercise but "innercise."

Massage

Massage is another non-exercise practice that can be used on a therapeutic basis to stimulate and cleanse the organs of digestion. Regular weekly or biweekly massages, using Swedish or accupressure techniques, should concentrate on the digestive system. Therapeutic massage gets blood and lymph flowing, loosens blocks, both physical and bio-electrical and increases your "firepower!" If massage can loosen up stiff muscles, think of what it can do to relieve an overburdened liver or an over stressed colon.

Herbs

Charcoal	Caraway
Cinnamon	Fennel
Celery	Cardamon
Clove	Anise
Ginger	Vinegar/honey
Cayenne	

Several herbs are beneficial to digestion. Cinnamon and cloves are stimulants that can be taken before a meal by chewing on a cinnamon stick or clove. A drink made from equal parts of honey and apple cider vinegar, diluted with water to taste, helps cleanse the stomach and prepare it for a protein meal. Cayenne pepper is good with a meal to help maintain temperature and circulation. Ginger root stimulates the salivary glands when chewed and helps reduce gas and fermentation when used as a tea. Cardamon, anise, fennel, celery and caraway are all good seeds to chew on after the meal to stimulate the flow of digestive juices, prevent fermentation, cramps and reduce gas. Charcoal is nature's finest absorbent of gases for the symptomatic relief of flatulence. It is made from coconut and is completely non-toxic.

Juices

Fresh fruit and vegetable juices before the meal can also aid in preparing the stomach. Sweet "live" juices, such as carrot, stimulate the appetite while green juices such as spinach or wheatgrass can quiet it. A juice that is too sweet, however, can turn off the appetite center. Children who are poor eaters at dinner time, should not be served juice first. Sweet, pasteurized juice such as common apple juice, creates a satiety because of its sugar. If you drink bottled fruit juice, serve it last. You may find your child's appetite will improve. A nutritious live juice, on the other hand, can satisfy you for 20 to 30 minutes because of its pure nourishment.

Enzyme Supplementation

There are many digestive aids on the market that support digestion. The most popular of these is hydrochloric acid. HCL is available in every drug store because of its common need by senior citizens whose production of HCL is naturally reduced. It exists in two forms, "betaine" HCL, derived from beets and "glutamic" HCL, derived from grains. Some nutritionists believe that betaine is most effective in digesting animal proteins and glutamic is best for grains and vegetable proteins. Whichever HCL you use, it will most likely be accompanied by pepsin, the enzyme which has the primary responsibility for breaking down protein. Pepsin is dependent on HCL because it can only function in a low PH range of 2.0 to 5.5. PH, is the measure of acidity and has a scale of 1 to 14 where 7 is neutral and numbers above 7 are alkaline.

Hydrochloric acid destroys harmful bacteria, which is a plus when traveling and consuming foreign diets. It has saved many Mexican vacationers from the infamous "Montezuma's" revenge. In fact, HCL is our best defense against auto-intoxication from impaired digestion. Here is an example of the power of hydrochloric acid. A few drops of the deadly toxin methyl guanidine causes convulsions and, in larger doses, death. Yet, according to Walter B. Guy, MD, it is made harmless in the presence of HCL. Professor A.E. Austin, in his book the *Manual of Clinical Chemistry*, tells us just how important the germicidal power of HCL can be:

> *When the hydrochloric acid content of the gastric*
> *fluid is deficient or absent, grave results must gradually*
> *and inevitably appear in the human metabolism. First of*
> *all, we shall have an increasing and gradual starvation*
> *of the mineral elements in the food supply. The food will*
> *be incompletely digested and failure of assimilation must*
> *occur. Secondly, a septic (pus forming toxins in the*
> *blood or tissues) process of the tissues will appear, pyorr-*
> *hea, dyspepsia, nephritis, appendicitis, boils, abscesses,*
> *pneumonia, etc., will become increasingly manifest.*
> *Again, a normal gastric fluid demands activity of the*
> *gall bladder contents and the pancreas for neutraliza-*
> *tion. Deficiency of normal acid leads to stagnation of*
> *these organs, causing diabetes and gallstones. In other*
> *words, an absence or a great deficiency of HCL gives*
> *rise to multitudinous degenerative reactions and prepares*
> *the way to all forms of degenerative disease.*

A good digestive enzyme supplement will also attempt to restimulate natural production of HCL in addition to supplementing it. Potassium and ammonium chloride aid in acidifying the intestinal tract and getting at the cause of the lack of HCL. These two minerals control chloride ion concentrations and regulate intestinal function.

One of America's greatest dietary complaints is over acidity. Ironically, it is often caused by the lack of hydrochloric acid. When HCL is insufficient, food ferments releasing unhealthful fermentation acids. These acids can cause ulcers and gall bladder attacks. Taking anti-acid tablets or liquids, as Americans often do, neutralizes the acids of fermentation, but makes the stomach too alkaline for the normal digestion of food.

Common Digestive Supplements

Papain	Bromelain
Lipase	Bile
Pepsin	Pancreatin
Protease	Glutamic HCL
Amylase	Betaine HCL
Lactase	

Over-the-counter digestive supplements contain other enzymes which serve various aspects of the digestive process. The papaya fruit has long had a reputation for its protein digesting enzyme papain. Papain is found in the unripe papaya fruit and leaves. Natives who live in the subtropics where papaya is grown, use the leaves as a meat tenderizer. They slit the leaves to allow the milky juice to extrude and then wrap the meat in them. The papain starts digesting meat protein thus increasing its tenderness. Bromelain, derived from pineapples, is another tropical fruit enzyme that helps break proteins apart into amino acids. Medical doctors have also used bromelain to clean out arteries and reduce inflammation.

Fat is digested by lipase and bile. The latter is often derived from oxen and would not be acceptable to vegetarians. Fat digesting enzymes can improve constipation by stimulating the flow of bile and improving the efficiency of the gall bladder.

Protein is digested by protease, trypsin, HCL and pancreatin. The pancreas secrets a range of enzymes, pancreatin, that work on protein, starch, cholesterol, RNA and DNA. All proteases, or protein digestive enzymes, split proteins into peptides and amino acids.

Amylase is also secreted by the pancreas, as well as the salivary glands and has the primary responsibility for the digestion of starch. Starches are carbohydrates which includes sugars. Many sugars have their own specific enzymes. Sucrose, lactose and maltose, for example, are digested by the enzymes sucrase, lactase and maltase. Fiber, on the other hand, is a complex carbohydrate with so many chemical links (polysaccharide) that it is virtually indigestible. Cellulose, the most prevalent carbohydrate in nature, is non-digestible. Its function is to contribute bulk to the intestinal contents.

Digestive aids serve a role on a therapeutic basis to help reclaim digestive strength after an illness, a fast or whenever the digestive system is weak. When used in this fashion, it is a bridge to help you get back on your feet. Some people's digestion is so weak that they absolutely depend on it. Diets high in overcooked vegetables, fried foods, canned and frozen foods are very lacking in enzymes and can benefit from supplementation. The enzymes in raw foods help in the digestion of the foods themselves. Enzymes perform many functions in the body including the repair and removal of old or diseased tissue. The main argument against exogenous enzymes is that they make our own enzyme-producing organs, lazy. For this reason, supplements should only be used on a short term therapeutic program or under the guidance of a health professional. If you think you are becoming too dependent on digestive aids, wean yourself from them. Unless you are on a therapy program, use these aids only periodically or whenever indiscretion turns into overindulgence.

Colon Drink & Cleanser

It's simple. Blend 1/8 cup of flax seeds for 60 seconds or until fractionated. Add 1 cup of apple juice and 1 small banana. Blend until smooth. It should have the consistency of a milk shake and will thicken the longer it sits. Drink before it thickens and follow with plenty of water or juice. Take on an empty stomach and eat no sooner than 1 hour later.

Flax has the ability to function as a bulk laxative because of its gelatinous nature. It moves through the intestinal track like a broom sweeping everything in its path. Follow this drink with plenty of water to keep the mass fluid. Both chia and psyllium seed are also gelatinous and have the same effect. Try this recipe with chia instead of flax or mix both together.

Be Happy

It sounds facetious but the same person eating the same meal can have vastly different results depending on his/her mood. A perfectly combined meal eaten while depressed can turn into poison. On the other hand, an enormously overdone feast can turn out well if there is a positive attitude, good company and good cheer. *(See p. 90 for more on happiness and digestion.)*

The body and the mind are interconnected and interdependent. The body expresses the thoughts of the mind. Constantly thinking crooked thoughts will create a crooked body. If you have a happy mind, your face and body will reflect that happiness. Everybody will know something beautiful is happening within you.
                                                    Swami Satchidananda

How To Control Appetite

Controlling how much you eat is one of the secrets to good digestion. Of course you could discipline yourself by preparing your plate with a pre-determined, fixed amount of food or by simply not taking a second helping. But the ultimate challenge is to stop eating when you are half-full. It may sound overly simplistic, but with this simple act alone, all of your digestive troubles would be solved. Another trick is to get up from the table and brush your teeth. You simply won't go back to eating. It works every time.

Certain foods also control appetite. Green juices such as wheatgrass, Green Magma, Chlorella, spinach and parsley create a satiety because of their concentrated nutrition and protein. Psyllium seed makes a blended drink that creates a feeling of fullness due to its bulk fiber. If you have a problem controlling your appetite, "drink" green and carry a toothbrush.

Factors that Weaken Digestion

Avoid eating while...

Talking on the telephone	*Walking*
Reading the newspaper	*Watching T.V.*
Riding an elevator	*Standing Up*
Driving the car	*Lying down*

We take the act of eating so much for granted that we rarely pay attention to it. When reading the newspaper, most people finish eating when they finish reading. If munching when on the phone, they stop munching upon hanging up. The talking or reading is the overriding activity...not the eating. Talking also forces air to be swallowed with the food and usually shortens the degree of mastication. Walking is another wonderful activity, but not while eating. Food should be eaten sitting down.

Parties

Food is entertainment and entertainment usually includes food. But that does not mean that your parties should lose all connection to nourishment and good food sense. Just because most parties serve potato chips and pretzels does not mean there are no healthy alternatives. Use the party as an opportunity to introduce your friends to new and healthier foods. Food creates social interaction. Let them become interested in new food ideas and tastes and they will learn about a new side of you in the process. Don't sell them on the idea of "health food," just new food. In this way, your friends will support your effort to eat better and learn something in the process.

Status is sometimes measured through food. Foods can be fashionable and even an indication of social position. Often, they are presented like decorations, an extension of the interior design. Advertising makes food stylish and lures us to various packaged, processed delights and even dares us (*"Bet you can't eat just one"* commercial.). What you eat can mean the difference of whether or not you fit in. The conversation may drift to your brand of caviar, your selection of fine wines, or choice of hors d'oeuvres. But food knows nothing of status. It is just food. Fashions are better expressed through clothes. In this convoluted world, food is used for pleasure, entertainment, emotional support and, on rare occasions, nourishment. By all means enjoy your food and eat foods that are fun. But do not depend on food for a good time or psychological support when you are depressed.

Business Lunch

By the same token, food should not be a tool of business (unless you are in the food business). Business needs your attention. Don't be put into a compromising condition by an escargot or a vodka sherry. You are obviously being sweetened for the sale. Or you are fattened with so much creme de la creme that you can't think straight. Alcohol is a wonderful negotiating tool. On the other hand, business is not good for digestion either. Stress over how much money is involved or the tension over whether a deal is made or lost is enough to turn your good meal sour--in your stomach. Make your deal first, then relax and enjoy your meal.

The Six O'Clock Fight

For many families, mealtime is the only opportunity they have to convene during the day. Consequently, it is often devoted to family "business" and this can lead to discussion or disagreement. In this atmosphere, it is not uncommon to embolden your argument with some flying spaghetti. Or you may cool down your opponent with some soda pop. In either case, your digestion suffers. Anger tightens the ducts and glands and reduces the secretion of hormones, enzymes and digestive juices. The nervous system which is in charge of sending the proper signals to the organs of digestion, is overridden by the emotions of the moment. You will literally loose your appetite and turn off your stomach creating gas and excess acid. On the other hand, good company and stimulating conversation are the most supportive conditions for good digestion.

(See also p. 23)

Interview with Swami

SPROUTANANDA

Swami Sproutananda is the guru of all sprouts, the origi-
nator of the food combining theory and chief authority of
all things comestible. Here, he shares his wisdom on the sub-
ject of digestion received by postless transmission from his
retreat at the top of Mount Germarest in the Himalayas.

Dear Swami, tell us about your diet.
I have tried every diet possible. I have been a vegetarian for
90 years, a strict raw foodist for 40 years and a macrobiotic for
25 years. I was a fruitarian for 12 years during my stay in the
Lesser Antilles and ate nothing but ice cream for 6 months in
the United States.

Why did you eat ice cream?
Ice cream is a happy food. People eat it to make themselves
feel good. And sometimes they don't feel so good, so they eat
it to feel better. I wanted to learn if this food has the intrinsic
capacity to make people happy. Is it the vitamins; is it the
enzymes, the RNA & DNA?

What did you learn?
First of all, I learned that there are many different kinds of
ice creams. I don't mean just the flavours or the colors but also
the very concept of what ice cream should be. I learned that
on an ice cream diet, one should live in America, preferably
Disneyland. Here in Tibet, we have only two flavors, cow and
goat. But in America, you have Baskin and Robbin. I could eat

all day at Baskin and Robbin. The next day I would spend at Ben & Jerry's. There, I not only experienced different flavors but also the third world, Vermont and the Rainforest all through the ice cream. In America, ice cream is a whole meal. You can have sandwiches, waffles, yoghurt, cookies, raisins, nuts, sprinkles and M&M's. I felt I was breaking my diet, but grateful that at least everything was vegetarian.

Did it make you happy?
For the most part, yes. Ice cream has the capacity to tickle your insides. But it has nothing to do with the ingredients. You see, ice cream puts you into an altered state. When you eat it, you think you are being given a treat. In truth, ice cream has no power itself and is neither good nor evil. Food, in general, has no morality. You may feel sinful when you eat devil's food cake because it is so delicious. But there are no devils around it nor are there angels around angel hair pasta. Candy bars do not conspire to rot your teeth. It is your belief system that makes these foods good or bad. That is why I experimented with the ice cream. If I had been taught from childhood that ice cream was the mainstay of my diet, then it would have no special significance. It would just be ordinary food and I would be looking for something else to make me "feel" happy. So, after 6 months, ice cream was just ordinary food.

What did you eat after the ice cream diet?
Nothing. I fasted for 2 years.

What foods make you happy now?
This is precisely what I learned on the ice cream diet: Happiness, is not a function of food. Happiness is an inner nourishment. You bring your happiness to the dinner table. The dinner does not make you happy. However, your relationship with food can support or disrupt your emotional state. If you are at peace when you begin eating and your meal causes indigestion, then you have disturbed that peace, your happiness. Food does not create happiness, but it can disturb it or support it. That is

why we must carefully select what we put inside our bodies and also be careful about the manner of delivery. Eating too quickly makes poison out of the healthiest, most nutritious food. Eating when you are not hungry is like forcing yourself to drive a car when you are tired. If you are miserable while you are eating, you are not prolonging your life, you are prolonging your misery. Choose a diet that both supports your body and your happiness.

Design your eating environment. Decide: Do I want to dine alone? With Friends? In a restaurant? In silence? Indoors? Outdoors? What gives you pleasure? Enjoying the process of eating is as important as selecting the food. If you are enjoying your meal, you are assimilating every vitamin and boosting your immune system. Scientists would have us believe that the immune system is supported purely by nutrients. But it is not. It is nourished by positive energy. When I was on my 2 year fast, I had no outside source of nutrients. Yet, my immune system was as strong as ever. Of course, food will contribute its nutrients, but it is your enjoyment of the food and positive feelings around it that send the strongest message to the immune center.

What do you eat now?
At my age, I have eaten everything. Now, I take enormous pleasure in just breathing. But how does it serve you to know what an old man eats? The body changes with the different ages. What an infant eats is different from what a child eats is different from what a teenager eats. And an adult would never eat like a teenager. So too, grandparents eat differently than parents. We must be in tune with the process of change. I ate only raw foods for 40 years. It served me well, but that phase is over now. My body has benefited and achieved what it needed from that diet. You are wearing a coat here on top of the mountain, but you will have to shed that coat when you descend to the warmth of the valley. You must change because your situation has changed. You must shed your coat though it served you well and you are grateful for it.

Asking me what I eat is simply a curiosity. It will not help you find out what you like. But to satisfy the curiosity I will tell you how I choose the food I eat. Most of all, I am nourished by love. I eat whatever people serve me when it is served with love. I have visited the fast food restaurants in America where the kitchen is mechanized and the food is prepared on an assembly-line, shipped by truck, microwaved and served. There is no comparison between this and food prepared by conscious chefs who take pride in their art and enjoyment in the feeding of others. There is a difference in vegetables grown by farmers who respect the land and nourish the soil by natural means; who garden according to the rhythms of the earth. There is a difference when a chef chooses quality in all the ingredients. Even if he/she puts a little too much sugar, it cannot hurt when it is prepared with such love.

How do we know if our diet is fulfilling our nutritional needs?

This is a big problem...not in getting our nutritional needs, that is easy. The problem is our knowledge is getting in the way. Modern science has provided us with wonderful information about food. Now we know all about vitamins and proteins and molecules. Did you know that no one knew about vitamins 100 years ago. How did they figure out what to eat! For thousands of years, people ate without knowing anything about vitamins! Today, we have charts and books defining the vitamins in every food and every item on the shelves of your American supermarkets has its nutrients listed and quantified. People even go to nutritionists to learn how to nourish themselves. They buy a shelf full of supplements in their attempt to juggle various nutrients into the correct combination for perfect health. Even with our modern scientific knowledge, we do not have the capacity to orchestrate the thousands of chemicals in food into the correct arrangement for health. You throw in a little vitamin A here, vitamin C there, a little B1, B2, B12, E, K, F, lysine, tryptophane. Do you really know what you are doing? Be honest. It is like playing Russian roulette. You know, a symphony may be composed of one million notes, but it has only 12 different tones. These 12 tones are played in different combinations and speeds and repetitions and punctuations which, in the hands of an artist, become beautiful music. Well, imagine having those 12 notes in your cupboard. You know how beautiful they can be, so you take lots of them and attempt to put them in order. But a symphony is not a scientific order. No knowledge of the specific tones and their effects will enable you to replicate the beautiful music. It cannot be done with the conscious mind.

So it is with food. Man should not attempt to second guess nature and that includes the natural forces inside our bodies. Can you orchestrate the type and amount of chemicals inside your body? That job should be left to your body and its innate intelligence. The little seed knows how to grow into a big plant. You may attempt to point its roots toward the ground or tug on its leaves to make it grow, but you would be better off just feeding it water. With your body, the best you can do is become a good listener. The more closely you listen, the better you will flow with the music. Tune into your body, not with a microscope, but with your eyes closed. Listen and feel your stomach. Connect with your food before your eat it.

People are looking to obtain health through diet. But if health is a rainbow, then diet is only one of the colors. To many people, health means "I am not in pain." But health is more than the absence of disease. It is an attitude. He was at ease. He dis-turbed that ease. Now, he is dis-eased. If you are looking to get healthy from food, you must develop a healthy relationship with food and a healthy attitude in your life. There is no perfect diet. When you sit down to eat, half the meal is the food and the other half is the feelings you bring to it. If you are fearful while you are eating, then you are eating fear. If you are miserable, you are prolonging your misery. This is so even if your plate is filled with the finest vitamin rich foods.

So, attitude is another color of the rainbow. You can eat the best diet and still be miserable. In fact, some people fuss over their diet so much, it adds to their misery. You may die early in life because of your misery and despite your good diet. On the other hand, you probably know of someone who smokes, eats pork, drinks and is happy at 85. Doctors have proven that smoking kills. But some smokers are alive and well at 90. Others die at 55. Other factors come into play. What is your attitude when you smoke? Do you smoke in response to stress or for enjoyment? Attitude is your salvation. That is why the "placebo" cures so many people. It is just made of sugar, but the patients believe it can cure them, so it does. What you think and feel has a greater effect on what you eat than the food itself.

You mentioned vegetarianism before. Is it the best diet?
We do not wish to kill any conscious beings. That is violence and violence makes the world a harsh place. Go to a zoo. Look at the animals. The carnivores are prowling and growling. They are restless and angry. Now look at the sheep, the cows, the horses and elephants. They sit in peace and eat from your hand.

As soon as you kill something it decays. Decaying matter creates poisons in your body. You have to work hard to counteract these poisons. Flesh also takes more time to digest. By the time you spend the energy to digest meat and counteract the poisons, you have spent more energy than you have gained from it. But a vegetable is easy to digest. When you eat vegetables and fruits, you are eating matter that is still alive. The cells contain live matter that you can transfer to your cells. Some people say "you are also killing the vegetables." But it is not the same. The vegetable kingdom is here to serve us. When you pick a fruit, the tree remains unharmed and ready to give more. When you harvest beans, some seed falls back into the ground and within 90 days, you have more beans. You can eat half a potato, put the other half back into the ground and get more potatoes. Try that with an animal and tell me what you get.

Many people in America are overweight. What is your advice for overeaters?

Almost everyone eats more food than they need. Fast, and you will be amazed at your ability to function without food. A big part of eating is ritual. If your ritual is three times per day, then you become accustomed to that. Here, on top of the mountain, the monks take only one meal per day. Their level of physical activity is low and the more quiet the body, the more clear the mind. We have no track runners or weight lifters here. Everybody has to regulate their consumption of food according to their physical requirements. But in Western civilization, people eat for the wrong reasons. If you eat out of boredom or depression, that food serves no bodily function. So it piles up. It's a simple law: burn it, or store it. Why do you think so many people have bad breath? Food, eaten in excess of the body's capacity, remains in the stomach undigested. Excess eating converts our stomachs into latrines. No wonder it fouls the breath. We are carrying a latrine around inside us! We have become a society of gluttons driven by our tongues, seeking evanescent sensations to fill the void created by our unhappiness. This misdirected search for happiness through food creates insanity. Some even take purgatives in order to vomit what they have eaten so they may be able to devour more. Yes, food imparts pleasure. But that pleasure comes from the satisfaction of true hunger.

Because of our bad habits and artificial way of living, few people eat for the purpose of nourishing the body. That is why so many are overweight. They view food as entertainment or use it to fill gaps of psychological need. Ninety-nine percent of the people who lose weight on a diet, gain it back again within a year. Although they succeed on the physical level, the weight problem returns because they make no other changes. If you question the successful dieters, you will learn that the only possible way to lose weight, is to gain life. The successful dieter launches his program with the goal: "I want to feel better about myself, I want to be happier."

Happiness should be your goal. If not, then why seek health? Is health to be sought after purely so we may indulge in bodily pleasures and take pride in bodily appearance? If this is an end in itself, then it is no better than drinking and bingeing and gluttony. We are caught in a struggle with Satan over control of our body. If we lose, then our body becomes nothing more than a filthy vessel with poisons and odors and anger oozing from every orifice and pore. Instead, strive to make your physical body an abode of God. Our body is a gift that has been given to us in purity and perfection at birth. It is our duty to keep it pure, both from within and without, so that when the time comes, we may return it in the same condition we received it. We cannot return exactly what we take, but we can convert the food, air and water we take into positive energy and use it to benefit the world. Then we are not debtors.

BIBLIOGRAPHY

The New Vegetarian, by Gary Null. Delta, 1978 New York.

Composition and Facts About Foods, by Ford Heritage. 1968, 1971. Health Research, PO Box 70, Mokelumne Hill, California 95245.

Food Combining Made Easy, by Herbert M. Shelton. School of Natural Hygiene, 1951 San Antonio, TX.

The Oil Story, by Eva Graf. Center of the Light, 1981. Great Barrington, MA 01230.

Nutritive Value of Foods, U.S. Department of Agriculture Consumer and Food Economics Institute, Agricultural Research Service, 1971. Washington, D.C.

Encyclopedia of Medicinal Herbs, by Joseph Dadans, N.D., PhD. Arco Publishing Co. 1975, New York, N.Y.

Back To Eden, by Jethro Kloss. Woodbridge Press, 1939. Santa Barbara, CA.

Dictionary Of Man's Foods, by William L. Esser. Natural Hygiene Press, 1972. Chicago, IL.

Toxicants Occurring Naturally in Foods, Strong, Frank M. National Academy of Science, 1973, pp.487-489. Chrmn, Subcommitte on Naturally Occurring Toxicants in Foods.

Food For Health, Ensminger, A.H., Ensminger, M.E., Konlande, J.E., Robson, J.R.K. Pegus Press, 1986. Clovis, California.

The Health Guide, by Mahatma Gandhi. The Crossing Press, Trumansburg, NY. 1965.

Nourishing Wisdom by Marc David. Published by Bell Tower, Stockbridge, Mass.

The Positive Times Newsletter. by Posner, Jerry, Wescott, Don & Jacquelyn. PO Box 244, W. Stockbridge, MA 01266-0244. Subscription $10.00 per year.

Beyond Words, by Swami Satchidananda. Holt, Reinhart and Winston, New York. 1977.

Nutrition, by Rudolf Hauschka. Rudolf Steiner, Press. 1967.

RESOURCES

American Natural Hygiene Society, Jim Lennon, Pres. PO Box 30630, Tampa, FL 33630. 813-855-6607

Natural Hygiene Society, 14 Lynn Haven Road, Toronto, ONT M6A 2K8. Gladys Arron, Joe Arron. 416-781-0359

Ann Wigmore Foundation, 196 Commonwealth Ave, Boston, MA 02116. 617-267-9424. Classes, clinic. The birthplace of wheatgrass. Full training and healing program.

Ann Wigmore Foundation, PO Box 429, Rincon, PR 00743, 809-868-6307. Puerto Rico location. Clinic. Full healing program.

Hippocrates Health Institute, 1443 Palmdale Court, West Palm Beach, FL 33411. 305-471-8876. 800-842-2125. Spa, clinic, wheatgrass training and healing program, classes, products.

Optimal Health Institute of San Diego, 6970 Central Ave., Lemon Grove, CA 82045. 619-464-3346. Hippocrates wheatgrass and training program. Clinic, spa, classes, juicers.

Creative Health Institute, 918 Union City Road, Union City, MI 49094. 517-278-6260. Clinic, classes, books, supplies.

San Francisco Living Foods Support Group, 662 29th Ave, San Francisco, CA 94121. 415-777-6874. Sprout line 415-751-2806. Live-foods newsletter, classes, dinners, networking.

Vegetarian Resource Group, PO BOX 1463, Baltimore, MD 21203. 301-366-8343. Newsletter.

INDEX

A

absorption....1, 21, 27, 29
acid foods....47, 59, 69, 73
acid, stomach....30, 33, 40,
      46, 47, 49, 51
acidity, excessive....82, 87
Advertising....5
adzuki beans....37, 50
Aerobic exercises....64, 77-79
affirmations for diet....7
alcohol....53, 74, 75, 87
alfalfa sprout....65
Alka-Seltzer....7
alkaline foods....1, 51, 82
allergies....11, 30, 32, 36, 37,
      52, 63, 65, 66
almond butter....38
almond milk....31
Almonds....31, 38, 46, 58
amaranth....36, 37
amino acids....45-47, 61, 65,
      82, 83
amylase....49, 51, 53, 82, 83
anger affecting
  digestion....24, 87, 96

Anise....79, 80
antacids....7
anti-acid....82
appetite....7, 18, 19, 53, 58,
      80, 85, 87
apples....29, 32, 33, 51, 59
arginine....61
aroma....17, 18
Art Of Food Combining....1
asparagus....32, 34
aspergillus niger....63
assimilation....19, 24, 27, 67
attitude....1, 78, 84, 93, 94
avocados....11, 33, 38, 58-60
awareness....19, 25, 63

B

B-vitamins....53, 75
bacteria....37, 40, 61, 63, 81
bacteria, cultures....63
baking....35
bananas....31-33, 70
barley....36, 50, 64

barley malt....52
beans....27, 30, 32, 36, 37,
            45, 50, 61-63,
            66, 94
beets....34
berries....51
Bibliography....97
bitter....18, 39, 49
bloating....28
bloodstream....49
brazil nuts....38
breadmaking....36
breakfast....47
breath....62, 79
breathing exercises....64, 79,
            95
broccoli....32, 34, 60
Bromelain....82
broths, vegetable....30-32
buckwheat....35, 36, 50
burping....24, 66
business lunch....24, 86, 87
butter....11, 58, 67-69, 74
buttermilk....40

C

Cabbage....34, 59, 63
cabbage family....61
candida....3
Candy bars....90
caraway....80
carbohydrates....25, 29-30,
            40-42, 48, 52,
            61
Cardamon....80
Carrots....34, 59, 80
cashew milk....32

cashews....32
cauliflower....34
Cayenne....79, 80
celery....80
cellulose....48, 49, 60, 83
cereal....73
charcoal....46, 64, 65, 79, 80
chard....34
Chart, food combining....54
cheddar....40-42
cheese....63
chemical combinations....1, 2
chemistry....1, 43-48, 67
chewing....8, 20-21, 24, 34,
            51, 53
    see also mastication
chia....65, 84
Chlorella....45, 65, 85
chlorophyll....58, 65
cholesterol....58
Cinnamon....80
citrus....4, 29, 30, 33, 47, 51
cloves....80
coconut....59
coconuts....33, 38
coffee....73, 74
colds....11
colon....47, 62, 64-66, 79
    cleansing drink for....65, 84
colors....35
complementing proteins....46
conversation....24, 87
cookie....68
cooking....25, 34, 35, 37, 42,
            46, 58, 60,
            62, 72, 83
corn....36

cottage cheese....40
cottonseed....58
Counting....24
cramps....80
cucumbers....33
cultured milk....40

D

Dairy....40, 63
dates....33
degassing beans....62
Di-Gel....7
diarrhea....4
Diary for diet....6, 7
Diet and aging....91
Diet vs. happiness....89-91, 94, 95
dieting....95
digestion....1, 6, 13, 15, 27, 34, 39, 43, 47, 51, 57, 58, 60, 61, 68, 70, 75, 78, 79, 87
digestion times....36
digestion, increasing....64, 77
digestive juices....5
Dinner....71, 74
discipline....6, 7, 20, 85
distension....12, 20, 21, 24, 53, 64
distress, gastric....9
dressings....34
Dried fruits....33, 39, 60, 65
Driving the car....85
Dulse....35
dyspepsia....24

E

eating and hunger....91
eating habits....1, 2, 6, 7, 12, 67
eating consciously....15
eczema....3
Eggs....41, 47, 58, 73
emotions....1, 87
energy....25
enervated....26
entertainment....86
environment and eating....1, 22, 91
enzyme....46, 51, 68
Enzyme Supplementation....80
enzymes....1, 13, 30, 33, 40, 57, 59, 68, 83
esophagus....53
essential oils....36
excesses....19
exercises....66, 77-79

F

Factors Weakening Digestion....85
family fights....24
  see also anger
fat....7, 12, 37, 38, 40-42, 49, 56, 57, 59, 69, 74
fat/protein/dairy....75
fatigue....11, 12, 42
fatty acids....56, 58
fear....93

fennel....80
fermentation....53, 61, 63-64, 80
fermented milk....40
fiber....4, 33, 34, 42, 49, 59, 60, 64, 85
fibrous....49
figs....33
Fish....42
flatulence....4, 12, 21, 24, 42, 61-65
flax....65, 84
flour....32
Food Combining Chart....54
Frances Moore Lappe....44, 46
fresh air....22
fructose....49, 52, 61
fruits....4, 13, 33, 51, 55, 60
frustrations....24
Frying....35

G

galactose....52, 61
gallstones....58
garlic....11, 65, 71, 74
gas....11, 20, 28, 61, 63-65, 80, 87
germinating beans....62
ginger....71, 72, 79, 80
glucose....49, 52, 61
Gluten....36, 63, 64
Glutenous grains....36
gluttony....19
Goat milk....40
gossip....24
Grace....21, 25

grains....27, 34, 36, 45, 49, 50, 52, 55, 68, 71
grapes....18, 32, 51
green....30, 33, 35, 58
green juices....80
Green Magma....45, 85
green salad....34
green....34, 71
grocers....17

H

habits....1, 8, 15, 23, 67
happier....28
happiness and eating....84, 90, 91, 96
also
happy....2, 70, 89, 90
HCL....64, 80
headaches....4, 7, 11-12, 67
health, meaning of....93, 96
heartburn....11, 28, 53, 75
Herbal teas....31
Herbs....79
hiccups....24
Hippocrates Health Institute....99
Honey....52, 80
hormone....46
hormones....87
Hunger....18, 19, 95
hungry....10
hydrochloric acid....51, 46, 59, 64, 80, 81
hydrogenated....58
hypoglycemia....3

I

ice cream....40, 89, 90
immune system....91
Indians....12
Indigestion....2, 4, 6, 7, 19,
            73, 74, 90
intentions....6
Intestinal gas....62
intestines....13, 31, 39, 42, 49,
            53, 78
irritability....11
irritable bowel....3, 21

J

juice....31, 32, 47, 51, 53, 70,
       72
juicers....99
juices....10, 13, 30, 31, 46, 80

K

kale....34
kasha....50
kefir....40

L

lactase, enzyme....30, 63
lactose, sugar....30, 52, 63
Latin Americans....12
laxatives....7, 84
Leafy vegetables....34
legumes, *see beans*....38
lemon....34, 58
Lentils....32, 37
lifestyles....1, 8, 27
liquids....13, 31

liver....46, 47, 49, 52, 78
Lunchtime.....10, 71, 74
lungs....78-79
lysine....61

M

macadamia....38
Macrobiotic....26
maltose....49, 70
mangos....32
margarine....56
massage....78, 79
mastication....20, 21, 24, 49,
           85
meal time....10, 12, 19, 24,
          87
meat....42
mild....18
Milk....30, 40, 41, 45, 52, 58,
      63, 72, 74, 75
milk sugar....52, 63
   see lactose
millet....36, 50
mind, influence of....6
miso....11, 63
mono-diet....29
Moon....12
mouth....20, 48, 52, 73
mucous membrane....64
mung....37
mushrooms....11, 35
music....25

N

nervous stomach....21, 57
nervous system.....17, 47, 56, 87
non-starchy vegetables....55
nose, sense of smell....16-17
nut butters....38, 56, 68
nutrition and science....91-92
Nuts....37, 39, 46, 54

O

oats....36, 50, 64
oils....11, 34, 35, 37-39, 56-58, 60, 69, 71
oligosaccharides....61
olive oil....34, 58
onions....5, 11, 65
Order of eating....28
Overconsumption....7, 39
overeating....7, 20
overloading the stomach....13
overweight....7, 95

P

pancreas....61, 78
Papain....82
papayas....32, 82
parasites....3
parmesan....40
parsley....30, 85
Peanuts....38-39, 46, 67, 68
peanut butter....68
peas....37, 38
pecans....38, 75
pepsin....46, 47, 60, 81, 82
peristalsis....49

picnic....22
pignolia....46
pine nuts....38
pineapples....82
pistachios....46, 74
placebo....94
polysaccharides....61
portions, size....7, 75
potato.....35, 50, 58, 60, 69
potato chips....71, 86
pranayama....79
  see also breathing
prayer....21
process....58
Protein....1, 37, 40-43, 45-47, 50, 53, 60, 64, 67, 68
protein combinations....53
provolone....40
psyllium....65, 84, 85
ptyalin....49, 51, 53, 73
pumpkin....46
Pureés....31
putrefaction....42, 47

Q

quinoa....36

R

raffinose....6
  see also flatulence
raisins....29, 33, 69, 70, 90
  see also dried fruit
rancidity....39
raw....34, 36, 59

Raw milk....40
Reading the newspaper....85
Regularity....10, 11
rennin....63
RESOURCES....99
Restaurants....11, 16
rice....36, 50, 70
rice crackers....50
ricotta....40
ritual--meditation....21
Rolaids....7
rutabagas....35
rye....36, 64

S

salad....34, 59
Saliva....48, 53
Salivary Glands....48
salt....11
sandwiches....67, 71
sauerkraut....63
schedule....9
seaweed....35, 46
seeds....37, 46
self-control....6
self-destructiveness....6
senses, five....16
sequence....28
siesta....10, 12
sight, sense of....18
silence....23
skin....47
sleep....11
smelling....16-17
smoking....67, 94
Snack Time....13, 73
society....6

soups....18, 30, 32
sour cream....40
soy foods....31, 45
soybeans....31, 37, 57
spicy....18
spinach....32, 80, 85
Spirulina....45, 65
Sprouted wheat....36
Sprouting....36, 37, 39, 62
sprouts....30, 34-35, 37, 59,
                62, 70-71, 99
squash....34
Squashes....50
Stachyose....61
starches..34, 36, 37, 39, 45,
                49, 51, 52,
                67, 68, 71, 75
starchy....34, 50
steaming....35, 60
stomach ache....4, 12
stomachs...5, 34, 40, 46, 49,
                51-53, 57-59,
                61, 66, 68, 78
sub-acid fruits....55
sugars....33, 37, 48-49, 52, 53,
                60-61, 63, 67-
                70, 72-73, 75,
                92, 94
sulfur....65
sunflower....38, 46
supplements, flatulence....64
swallowing....8
Swimming....78
swiss....40
system....91

T

Tahini....34, 38
Taste....18
telephone....85
tempeh....37, 45, 63
temperature....61
temptation....6, 39
thanksgiving....21
thickeners....32
timing....1, 13, 34, 46, 61
toast....73
tobacco....4
Tofu....37, 45, 71
Tomatoes....33, 51, 71, 74
tongue....18, 47, 53
trypsin....37, 61-63, 83
TV....15, 22, 85

V

vegetables....
Vegetables....4, 25, 30, 33-35,
                42, 49, 50,
                58, 60, 72
vegetarianism....94
vinegar....34, 51, 80
vitamins....2, 25, 31, 50, 52,
            56-57, 74, 89,
            92

W

Walking....78, 85
walnuts....38
watching TV....85
water with meals....59
watermelon....33, 59
weight, losing it....95

wheat....36
wheatgrass....65, 80, 85, 99
Wigmore, Dr. Ann....99
work-aholic....9

Y

yams....35
yeast....45
Yin and Yang....26
yoga....78, 79
yoghurt....63
Yogis....23, 25